RFID: Improving the
Customer Experience

DISCARD

Practical Books for Smart Marketers from PMP

Now you can equip all your sales people with **RFID: Improving the Customer Experience.** It will help them introduce new RFID solutions to your existing customers and open the doors for new business development. You may also want to distribute the book to potential customers to help them understand the power of this technology and how it can be used in their industry.

A customized edition, with "*Compliments of Your Company Name*" on the cover is available with orders of 200 or more copies.

Call us toll-free at **888-787-8100** for quotes on quantity orders.

For more practical books for smart marketers, go to our website, **www.paramountbooks.com.**

RFID

Improving the Customer Experience

One-to-One Marketing in Real Time

MICKEY BRAZEAL

Paramount Market Publishing, Inc.

Paramount Market Publishing, Inc.
950 Danby Road, Suite 136
Ithaca, NY 14850
www.paramountbooks.com
Telephone: 607-275-8100; 888-787-8100 Facsimile: 607-275-8101

Publisher: James Madden
Editorial Director: Doris Walsh

This publication is designed to provide accurate and authoritative information in regard to the subject matter covered. It is sold with the understanding that the publisher is not engaged in rendering legal, accounting, or other professional services. If legal advice or other expert assistance is required, the services of a competent professional should be sought.

All trademarks and service marks are the property of their respective companies.

Cataloging in Publication Data available
ISBN 13: 978-0-9801745-3-3 | ISBN 10: 0-9801745-3-8

This book is dedicated to Wilma Crumley, who was my teacher.

Contents

CHAPTER 1

Introduction: A search engine for things

Do you remember life before search engines?

There were lots of things that you needed to know, and couldn't find out. Lots of things that would be nice to know but were too much trouble to find out. Times when you had to do things the hard way because you didn't know the easy way existed. Search engines have made a giant change in the way people live and work and get things done.

But a search engine won't search for *things*. Or for *people*. It won't tell you where your golf ball went. Or who's at the front door. Or what's in the attic, and where in the attic it is. Radio Frequency Identification will.

Radio Frequency Identification (RFID) is built around the idea that you can put cheap and tiny tags on physical objects—tags that contain a radio transceiver on a computer chip—and then use the tag to know where the object is.

Radio Frequency Identification is a search engine for things.

What would you do if you knew where everything is?

What would you do differently? Suppose that, as suddenly as Spiderman, you get this superpower: you can know exactly where things are. Whatever things you're interested in. Millions of things. Not just categories of things, but individual things: this particular screwdriver, that particular snow tire. People too—this particular customer, the moment he walks into your store. If the person has given you permission, you know he is there. How would you use that power?

Suppose you're the person who runs the supermarket. And suppose you could know instantly, constantly, magically, the location of every beefsteak and banana and razor blade in your supply chain. Some things you'd do pretty quickly, it seems to me. First, you'd keep stuff from walking out the door. There is at least $30 billion dollar's worth of "shrink" at retail every year. None of that would happen in your stores. You could take that cost out of your operation, and gain a huge competitive advantage by sharing the savings with your customers.

Second, you wouldn't order more merchandise until you needed it. There are many billions of dollars in "safety inventory" in the U.S. retail supply chain, taking up space and using up capital. If you could do "responsive restocking"—putting merchandise in the store at the moment when it's needed to replace stuff that someone bought, and if you could buy just enough to get that done, you would subtract another big chunk of your total costs, and either become the most profitable or the lowest cost supermarket in the country.

You certainly wouldn't make your customer stand in a checkout line and wait while you pick up and scan every item in her cart, one by one. You'd *know* what she picked up. You'd hand her a list, and put it on her credit card as she walked out the door.

Maybe you would ask your customer for permission to accompany her, unobtrusively, electronically, as she walks through your store, the way Amazon does when you move through its website. You'd see what she looks at and what she misses, see what she picks up and then puts back. And maybe, as Amazon does, you would make individualized offers based on individualized needs. Coming in the door, she would touch her loyalty card to a kiosk, which would print out a bunch of special-price offers tailored to her personal shopping list. Why put the same things on sale for everybody? Everybody doesn't want the same things.

In-store advertising could also be individualized. Instead of a sign that says "Cake mix: buy one, get one free," as your customer goes by, the sign would say: "Mickey's birthday is coming up. He loves chocolate. Half-price, just for you." As the next person goes by, it could switch to a different message.

Or maybe that's a terrible idea. The point is this: this power to know where everything is, is here today. RFID tags can track each individual thing in a store, in the back of the store, in a truck on the way to the store, in a warehouse, and so on. *If customers want personal recognition and individualized offers*, it can find them and reward them as well. Automatic identification of people and things is doable now, and its cost is falling as its use grows.

So it's time to figure out what we want to do with it.

That's what this book is about.

Some stuff needs tracking more than other stuff does

It makes more sense to track expensive things than inexpensive things. It makes more sense to track things that are fragile or perishable or easy to steal or might need to be fixed. It makes more sense to track things when there is a high cost if they stop working. The benefits of tracking may be larger in larger markets.[1] *But we can also use RFID tags to do a better job of taking care of the customer.*

We can eliminate some annoyances. We can offer some new services. We can recognize the customer at the point of sale, and make a personalized offer that might make that customer happier. We can turn the retail store from a mass medium to the one-to-one interaction that it was always meant to be.

The idea of Automatic Identification

Think of an RFID tag as a radio transmitter, a radio receiver, and a little chunk of computer processing power, all on a chip about the size of two capital letters in this book.

You can put that tag on just about anything: a pallet load of frozen peas in the back

How RFID works is described in more detail in an appendix at the end of this book. You might want to stop and read it now, or refer to it when you come to an application that interests you. It includes sources for further reading.

•

of your grocery store, or a single case with 24 bottles of shampoo, or a single package of razor blades on the shelf. You don't have to put it on the outside. It can be (in most cases) inside the middle box in a pallet load of boxes, or embedded in the pallet itself.

An RFID reader, a radio interrogator that you can mount on a wall or carry around with you, can read every tag that comes within range. Instantly. Constantly. All of them, simultaneously. It can know when those items come out of the backroom into the grocery store. It can know when they go into a customer's shopping cart. It can know when they go out the door.

It can know just the location of the package of razor blades, or it can know a whole bunch of data that someone decided to attach: a warranty, or which distribution center it came from, or anything else that might be useful.

Your RFID reader can dump that information to a computer middle-ware system that decides, according to pre-set rules, which of these pieces of information you need and which you can ignore.

You can do all this for not very much money. Right now, you can buy the simplest tags for about ten cents apiece if you're willing to buy a million at a time. (Tags with sensors cost more.) And prices are falling. When we get to a five-cent tag and a hundred-dollar reader, then it's probably practical to track most everything in the grocery store. Already it makes sense to track more expensive things like shirts and shoes and whiskey bottles.

Most important of all, you can do all this without using human labor. That alone makes RFID vastly cheaper than any existing way of keeping track of things.

When you get all this information, which you couldn't possibly collect before, you can have a precise, up-to-the-moment view of exactly what is happening in any process that you want to pay attention to. Instead of checking from time to time and noting the changes since last time, you can know *now*. RFID is a cost-cutter for real-time visibility.

Real-time visibility changes everything

It almost forces process improvement.[2] As a vendor, if you discover that a special display case full of breakfast food you put on sale yesterday is still stuck in the grocer's back room in a store in Aliquippa, first you'll give that store a call to ask them to bring your display onto the selling floor. Then you'll figure out how it happened and how to keep it from happening next time.

Information processing has mostly been about collecting historical information in batches, and combining all the batches in some sort of database. From time to time, you go back and analyze it and figure out what to do. But now, like the stock traders whose computer systems automatically buy or sell a stock when it reaches a pre-decided price, we have to learn how to build *event-based systems*. When Mrs. Hanrahan, who always buys the 25-pound bag of dog food, comes in the front door, the system checks to see that there's a bag of her brand on the shelf, and if there's not, it sends someone to get a new one out to the shelf before she gets there.

Reducing the cost of data collection lets you be more granular in the data you collect.

Procter & Gamble has done tests to try to figure out how to put a smart shelf in the cosmetics department of the drugstore that will make certain every different shade of lipstick is in the right place and in stock all the time. (Customers keep picking them up and putting them back in the wrong spot.)[3] Without RFID, that's just too much to bother with. With RFID, it is, or will be, a solved problem.

Another kind of change happens when your customer figures out that real-time information is available. Today people walk into a bookstore, and check on a self-service screen to see if a particular author has a new book out, and if it's in-store. Tomorrow, perhaps they'll check a specific item in a Target or Wal-Mart and see on the shopping-cart screen where it is in the store, what it costs compared with the alternatives, and whether it is the most ecologically desirable solution.

Way beyond bar codes

The last great leap into automatic identification was the bar code. It cut billions off the price of goods at retail, made retailers and brand marketers more efficient, and felt so powerful and terrifying that several states forbade its use for a few years (at considerable cost to their citizens). It was truly a giant step in improving the customer experience.

This is a bigger step. RFID can do a lot of things that bar codes can't. You need a clear line-of-sight to read a bar code, but radio waves don't need line of sight. So the product can be facing the wrong way. It can be in the dark. It can be outdoors in glaring sun. It can be as dirty as a pig's ear. (Lots of pigs' ears have RFID tags).

Bar codes have to be "oriented" to be read; they must be held at a particular angle to the reader. In practice, this means they have to be held by a person, so there's a time delay while each item in a series gets picked up and wiggled in front of the scanner. Having a human being pick up each item adds so much cost that many supply chains never scan packages at all, from the time they are manufactured until the time the customer checks them out of the store. Visibility would cost too much.

RFID tags are read automatically, simultaneously, when they come in range of the reader. They don't have to be oriented. They don't necessarily have to be close. They can be in the middle of a stack. That makes them faster, which helps in the supermarket checkout line and lots of other places. When the Shanghai Post Office went from bar codes to RFID tags for tracking express-mail bags, it picked up a 20 percent productivity gain just from faster reading.[4]

Bar codes identify a category. The tag says, "This is our basic blue long-sleeved men's shirt." RFID tags identify a *specific instance* of the category: "This is the third of seven, different, size 16/35, basic blue, long-sleeved men's shirts we received last Tuesday." Customers who buy shirts in a department store know how often they pick out a shirt and then find their size isn't on the shelf. The sales associate didn't bring out more from the back room because nobody knows that all the ones in that size have been sold. RFID tags will fix that.

Bar codes are unchangeable after they leave the factory. With RFID tags, you can add information in the field. You can even re-program the tag to do something different.

Bar codes can't carry much data: 80 bits is the maximum. Boeing attaches RFID tags to specific aircraft parts, putting lots of information right on the part. They have 10 *kilobit* tags.[5]

Bar codes are pretty easy to forge. The checkout clerk can't read them, and might not remember the price of an item. There are thieves who stick a new bar code over the old one, buy the product for a fraction of its price, and then resell it on eBay. It's very tough to forge an RFID tag, especially if it's encrypted.

There's even a cosmetic advantage. Package designers hate the big ugly spot a bar code makes, especially in luxury items or fashion goods. An embedded RFID tag won't mess up the design.

You can design RFID tags to be disabled at the point of sale, so they're unreadable outside the store.

Now add sensors

In addition to identifying an object, you can collect data about its condition. There are sensors-on-a-chip for temperature, pressure, acceleration, (Was the product dropped? From how high?), presence of particular chemicals or radiation, and many others. Tags can be designed to send an alert if a sensor detects a change that exceeds a pre-set parameter.

If a piece of meat on the supermarket shelf is not what it should be, you can know with a sensor on a tag, before the customer carries it off and blames you. You can also know if it went bad at the store or before you got it.

RFID re-organizes relationship marketing

Over the last decade, marketers have shifted from a focus on selling products or services to a focus on acquiring and retaining customer relationships.

There are many reasons for this. Brands are not as powerful as they once were, especially among younger customers. Mass media advertising has lost both reach and persuasive power compared with its glory days. Most customers of most enterprises have lots of choices, and product differences are small. In this situation, as Gamble says "Relationship is the only sustainable advantage."[6]

Today there are many business models that can produce a return on investment *only* if there is a significant amount of customer loyalty. And loyalty requires a relationship. "Owning a relationship," as Kasanoff puts it, "is more important than owning the assets required to produce products."[7]

Loyal customers behave differently. They spend more with a given company. They are less price-sensitive. They are more forgiving when a company disappoints. Their predictability makes the company more efficient. Best of all, they recommend a company or product to new customers. Customers you get from recommendations are more likely to become loyal than those acquired by marketing.

Customer Relationship Management (CRM) tries to unify marketing, sales, and customer service activities to acquire, retain, and grow relationships. A company tries to build a learning relationship, in which the customer shares information with it so that the company can provide more personal service. At the Ritz Carlton Hotel, if a guest eats the banana in her fruit bowl, the next day's fruit bowl has two bananas. A smart cell-phone provider will look at a customer's usage and his contract to see if there is a contract that might fit him better, and thus *keep him longer*, whether or not it brings in more revenue this month.

The more important this one-to-one perspective becomes, the greater is the role of interactive, data-collecting and data-distributing technology. Jack Welch of GE (quoted by Peppers and Rogers) put it this way: "We have only two sources of competitive advantage: the ability to learn more about our customers faster than the competition, and the ability to turn that learning into action faster than the competition."[8]

If you can know, the moment a customer crosses your threshold, who he is and what he prefers and how he wishes a particular service to be delivered, then maybe you can do a better job. If you can instantly consult a real-time picture of your own inventory and processes and capabilities, then maybe you can do a better job. If you can create event-driven systems so that a customer's action instantly triggers a customized and helpful response, then maybe you can do a better job.

Needless to say, it's not as easy as that. People change, and what I want today is not what I want tomorrow. People cherish their privacy, and place great restrictions on what they want you to know about them (though this is complex and changing). Nobody wants to tiptoe through the drugstore in fear of setting off the tripwire of another marketing barrage. I am here to meet my needs, not yours; leave me alone.

And yet . . .

Marketers have spent billions in various mass-marketing attempts to "build the brand relationship." In most cases, that's a one-way relationship. I know Rice Krispies, but they don't know me. Peppers and Rogers talk about building the *branded relationship:* a two-way process in which the brand is aware of the customer and constantly making little changes to meet the needs of that individual.[9] To make that happen, marketers will need the low-cost and real-time responsiveness of RFID.

Customer experience is key

Some critics of CRM believe it has already slid too far into a sort of bloodless optimization—that, in attempting to treat different customers differently, it treats them all as fields in a much-too-personal database, a parody of the offered relationship.

There is enormous danger in automating any part of the customer experience. Corporate call centers were invented to provide better customer service, but quickly became an object of derision:

> If you are deeply irritated by this recording, please press 1. If you notice that you've been on this call for too long already, please press 2. If you long to express your outrage to a responsible human being,

but know that it isn't going to happen, please press 3. If you feel your life sliding slowly into meaninglessness and despair, please press 4.

It is the fully automated expression of disdain for the customer.

The way out of this is to focus on the customer experience. This is where CRM started and where it must return. The thing to remember is that, in many categories, a customer's experience while buying a product is almost as important (or sometimes more important) than the product itself. Unfortunately, it's also sometimes harder to produce than the product itself. "Unlike low prices, unlike great products, great service is difficult to copy."[10]

Here is an important chance for RFID to add value. RFID can empower new services that make the experience faster, more convenient, more intelligent, more fulfilling, healthier, safer, more eco-friendly, and, in fact, more personal.

Enough of abstraction: bring on the applications.

Supply chain has paved the way

Most RFID users see it as supply chain technology, pure and simple. Tracking inventory from manufacturer to distribution center to warehouse to retailer's back room, all the way to the selling floor is critical enough all by itself to make the technology succeed. In addition, the vast volume of tags and readers required by supply chain applications will bring about the cost reductions necessary for RFID to be practical in customer-experience applications. The middleware that sifts billions of events to find just a few that must be attended to was developed for supply chain needs.

This book will not give much space to supply chain applications, because other books already have, but supply chain problems certainly affect the customer experience, so they are summarized here.

Fast-selling items are out-of-stock at retail about ten percent of the time.[11] RFID tags can move that number close to zero. As tags move out of the store, middleware keeps track of them. It can summon "responsive replenishment," not based on predictions of demand, but on actual demand in that store on that day. It's an enormous pain for a customer to drive all the way to Wal-Mart and discover that some of what he wants is not on the shelf. It's equally painful for Wal-Mart to lose the sale. Wal-Mart has been a global leader in solving this problem with RFID, essentially forcing its largest suppliers to tag their products, and thus jump-starting the whole industry.

For the retailer and the manufacturer, there's lots of money in managing inventory better. If retailers know where everything is, they don't order so much safety stock. They don't re-order a product, just because they can't find it.

Many of the suppliers who ship to retailers believe that the retailers' inventory records are often wrong. Garfinkel cites a study of a retail group in which 65 percent of all the inventory records are "significantly inaccurate."[12]

A classic problem in supply chain management is the "bullwhip" effect, in which little ups-and-downs of inventory at retail cause managers to make somewhat larger additions or cancellations one step up the supply chain. These ups-and-downs cause the distribution center to add orders or cancel orders from the manufacturer in still larger amounts. And those added orders or cancelled orders cause even larger ups-and-downs in components ordered by the manufacturer. A little volatility at the customer end of the supply chain becomes a company-destroying whiplash at the other end. But when inventory up and down the chain is made visible with RFID tags, the bullwhip effect disappears.[13]

Professor Bill Hardgrave at the University of Arkansas, who led and studied Wal-Mart's first efforts with RFID says, "We added one key piece of data: Do we have it in the backroom?" The impact, he says, was phenomenal. Items that sold 6 to 15 units a day showed a 62 percent decrease in out-of-stocks. And associates who were previously working on catch-up replenishment problems could focus on customer service instead.[14]

The retailer harvests the biggest gains from better supply chain management. Retailers have jumpstarted the process by requiring their largest suppliers to set up RFID tagging processes that fit the retailer's eco-system. Mandates for RFID tagging have come from Wal-Mart, Target, and Albertsons, from Tesco, and Metro in Europe, from the U.S. Department of Defense and others. Manufacturers whose relationships with these retailers are critical to survival adopted RFID tagging without too much attention to its benefits for their own firms. But now that it is in place, some are learning how real-time visibility can drive their own process improvement. Others will follow.

Supply chain is only the beginning. In the chapters that follow, we'll explore the other ways you can use RFID to transform the customer experience. New business processes, and whole new businesses are created by the ability to track assets cheaply and automatically. We can begin to stamp out counterfeiting in drugs and in branded products. It's not just Vuitton and Viagra—Procter & Gamble says it has found counterfeit Prell shampoo within a few miles of the home office. Hazards to the food supply, from terrorism to Mad Cow Disease can be prevented. New payment systems make it easier to buy things in a hurry, and that makes people buy more things. New kinds of tickets eliminate waiting in line. New, personalized retail experiences become possible. Healthcare becomes more caring and more efficient. Perhaps most important of all, our desire for a greener future and a sustainable society move, in some important ways, beyond aspiration to practical techniques. This book will also try to show that, properly controlled, RFID can meet our urgent needs for personal privacy and for control of personal identity.

The Internet of things

Look down the road. Techno-prophets describe a fast-approaching age of ubiquitous computing—an Internet of things. The idea here is that most computing will be done outside of computers, that many physical objects will have some kind of information processing attached to them, and that these objects will communicate with users and with each other, via the Internet.

This is a book about how marketers should plan to make use of technology that is available or becoming available now. It is not suggested that you need to revisit next year's promotion plan to take account of the Internet of things. But, to understand the impact of this new technology, and the urgency and certainty of its advocates, it is important to look ahead and see what others see.

Computers embedded in products will become more important than computers that stand alone. To a great extent, they will communicate with each other, for the benefit of people, but without involving people. If the shirt has processing power, and the washing machine has processing power, maybe they should talk. "Don't put me in with the underwear. I want warm water in the wash cycle and cold water in the rinse cycle. If you let it get too hot, I'll shut you off." Some IT experts think that application is silly; it's too much work for not enough benefit. But who knows? It might get easier. There might be people who care more about their laundry than IT experts do. On the Internet, things can talk to other things without regard for distance.

The Network Effect is the idea that the usefulness of information-sharing products increases exponentially with the number of nodes. When a thousand people have an iPod, it's not much use. When a hundred thousand people have an iPod, it makes sense to think about how to add to the music they can play. When ten million people have an iPod, you have to figure out how to redesign the business models of the music industry.

Today there are RFID tags on pallet loads of products in the backrooms of supermarkets. There's a lot you can do with that. But when there's a tag on every package of T-bones in the supermarket, what you can do as a marketer will change. The issues you must worry about as a marketer will change. You might as well figure it out before the other guy does.

Imagine that this "ubiquitous computing" is inevitable. Lots of well-informed worriers believe this is true. Everything that can be digital will be. Everything that can be linked to other things will be. Imagine that this happens quickly. In such a world, new kinds of information will be available in real time wherever you happen to be. Pick up a roll of paper towels and your shopping cart shows you the cost-per-sheet of the different alternatives on the shelf, and their environmental sustainability rating. Pick up a bottle of wine, and you get its CNFATE tasting score.[15]

The point is this. In a world where all sorts of technological capabilities, both known and unknown, are going to be available, the important task is to figure out *what to want*. Information technology people have not always been good at this. Our IT creators, revolutionaries though they were, seem to have wanted incomprehensible computers, not optimized for any particular task, physically separated from the devices they direct, run by bloated operating systems, which can respond to hundreds of thousands of instructions, if anyone could remember them all and know which instructions to give.[16]

But marketing people are all about figuring out what to want. As you read, in the following pages, about the first steps taken with RFID to make relationships more satisfying and improve the customer experience, think about how you will use the power to know where everything is, and what condition everything is in. *Think about what ought to be.*

Time's a-wasting.

CHAPTER 2

RFID and Relationship Marketing

HARRAH'S

Which kind of gambler are you?

Casinos are pioneers in relationship marketing, and Harrah's is the pioneer among casinos. Its customer-relationship database was up and running ten years ago. Its operations take their shape from the need to deliver an experience tailored to each customer's individual relationship, in real time. It measures success against sophisticated predictive models for time spent, share of spending, and retention.

Harrah's works with four tiers for customer value: Gold, Platinum, Diamond, and Seven Stars. A loyalty card tracks spending on gaming, restaurants, retail and entertainment. Inserted in a slot machine, it acknowledges every nickel. The more you spend, the more you get back. New customers get a Gold card, and work their way up. Rewards are lavish, and constant, and clearly differentiated. At every interaction, literally all day and all night, the customer sees special treatment based on that loyalty tier.

Promotional offers are different. Room rates are different. There are exclusive lounges for Diamond and Seven Star customers. There are elite concierge services. Each of the hyper-VIP Seven Star customers has an individual host on call around the clock. There's an active process to invent new perks and pamperings. Creative people spend serious time thinking up things to give away, ways to recognize and reward.

Seven Star and Diamond customers never stand in a line. Think about the operational issue there: two entry points, two different protocols at check-in, at restaurants, at shows, at cashier windows. But every customer sees what happens if you stay loyal to your favorite casino.

Casino brands are strongly segmented: the Harrah's experience is different from the Horseshoe experience is different from the Caesar's

Palace experience in look and feel, even though fundamental dimensions of the gaming product are identical by law. But those three properties and every other brand owned by the Harrah's group will instantly recognize and reward based on the loyalty points you earned at any Harrah's property.

Offers to customers are centrally managed, so the customer doesn't get an irritating deluge of promotions from everybody at the same time, and so that promotions fit the personality of the brand as well as the status of the customer.

The core benefit of the casino has something to do with spending a moment in the spotlight—living life on a larger stage. Recognizing the customer's special status is part of that—part of the product as well as the marketing. It is so central to the value of the business that casino properties today are bought and sold at a price based, not so much on their physical assets, as on the value of the portfolio of relationships they have grown.[1]

It is a basic premise of this book that marketers are moving from classical, brand-based mass marketing toward relationship marketing and customer-experience management.

In classical mass marketing, the customer relationship is with the brand. The brand is a promise about the quality of the experience a customer will have. It is the reservoir that captures awareness and preference created by marketing. The brand is built, differentiated, and maintained by mass communication. Communication is aimed at broad categories of people, perhaps defined by some demographic characteristic. It is not considered necessary for the marketer to know the identity of the individual prospect or customer. It is assumed that, at the point of purchase, customers will select the brand that has won them over with effective mass communication.

There is both truth and usefulness in these ideas. But the world has changed in ways that make it difficult for this mass-communication and brand approach to be as effective as it once was.

First, meaningful differences between brands have become hard to produce, or less important, or unreliable or unbelievable. Even where strong preferences exist, the alternatives are not that unacceptable. If, at the supermarket, the beer you love is not on sale, you may find it possible to love the one that is. All of the shampoos are good enough. No airline is good enough, and so on.

Second, the cost of building brand preference through media advertising has become almost unsustainable. Media prices have increased faster than inflation for decades. Media audiences have fragmented; a prime-time network TV show may deliver only three percent of the total audience. Competition from email, cell phones, social sites and the like has degraded the attention paid to TV in particular. Yet the amount of extra margin you can get for being the most-preferred brand may not have grown much at all. Basic, blue-chip brand advertising has become a high-risk investment.

It is made riskier still by the persuade-and-discard nature of the process. Advertisers must send their message to a broad, anonymous market segment, not knowing who has been exposed to it once and who has seen it a hundred times, not knowing who in that audience is a hard-core loyalist and who is a hard-core rejector (neither of whom needs more advertising). With each new message, the advertiser starts over, speaking blindly to people whose relationship with the brand is unknown.

The third assault comes from mass customization. Mass marketing assumes a mass product. Everybody who buys Wheaties gets the same stuff. Now however, it is technologically possible, and often cost-effective, to make a different product for different customers. Count the variations available in a new car. Francis Buttle counted 27 million potential combinations in the Ford Fiesta, in one model year.[2] It is mathematically possible for every customer to buy a different car. It probably won't happen, but it wouldn't change the numbers if they did. It's mostly a data processing problem, and data processing has become inexpensive. Online, you can buy a somewhat customized computer. You can have a customized teddy bear, a customized make-up blend, a customized investment plan, or customized blue jeans with no great cost penalty to the producer. Simple and widespread customization subverts the mass marketing model in two ways. First, it exalts feature collection at the expense of the brand, and second, you have to know the individual customer to get it done.

But the most important driver of divergence from brand and mass marketing is the spectacular fall in the cost of data processing and data collection. These days, even in some large-scale consumer businesses, it's possible to track the purchases of individual customers. Everyone who tries that discovers the same thing. *A small number of customers are so critically important to the success of a brand, that it is simply too dangerous not to follow what happens to them as individuals.*

The number of people who will predictably spend a lot of money with airlines is so small that, inside their "we love you" programs, they have "we really love you" programs, and "we love you even more than we love those other people" programs inside of those. The tactics feel fickle, but the goal is right. An airline must get and keep its share of high-volume flyers to remain profitable. A small percentage of gamblers account for most of the casino's profits. A few retail bank customers make the bank profitable, or not. In a Coopers and Lybrand study of retailers, the top 4 percent of customers produced 29 percent of profits. The bottom 70 percent produced 16 percent of profits.[3] Tesco says the top 100 customers in a supermarket generate as much revenue as the bottom 4,000.[4] There are certainly exceptions, but don't assume your business is one of them.

The basic idea in relationship marketing is to replace the blind "persuade and discard" process with an "acquisition, retention and development" model.

A firm must *acquire* potentially profitable customers. It costs less to make this happen with Internet offers, tightly focused on people who have already displayed a targeted behavior, or with offers at the point of sale, or rifle-shot database marketing. People who specialize in acquisition will tell you that it's not about the brand; it's about the offer. The brand must stay general, but offers can be individualized. This is not the same as acquiring product sales, or acquiring market share. The goal is the beginning of a relationship with an identified individual.

A firm must *retain* identified profitable customers. You know who they are, so it's possible to talk to them one by one, with at least a little reference to the existing relationship. Part of retention is winning back customers to whom you accidentally delivered a bad experience. An even bigger part is minimizing your expenditure on unprofitable customers. Mass advertisers spend a lot of money wooing customers who will cost them money.

A firm must *expand* or *develop* its relationship with profitable customers. Here again, the focus is on the identified individual. You try to upsell and cross-sell customers to get them to buy more profitable or different products than the ones they have already tried. You use the information that you get from the relationship to try to do a better job of paying attention to your best customers and creating the highest-quality experience possible for those customers, thereby strengthening the relationship.

Relationship marketers describe the process of understanding customers better, along three dimensions.

First, you want to understand *customer value*. If there's a way to invest more in some customers than in others, it ought to be the customers with the highest lifetime value. Supermarkets offer special lanes to give some customers faster service. The sign over those lanes says "Ten items or less" or "Fifteen items or less." In a rational world, the sign would say "$12,000 a year or more." To predict customer value, the marketer tries to quantify each of the factors that create value: cost of acquisition, length of retention, number of referrals, cross-sell and upsell. Relationship marketers will have a prediction of value added per year of retention, based on the characteristics of that particular customer. Customers that you get with one kind of offer might have a different retention potential than customers you get with a different offer, for example.

Second, you want to understand *customer preferences*. Is this customer all about price, or focused on convenience, or moved by your ecological advantages? Which values drive behavior, and which are just nice to have?

What trade-offs are desirable? There are people who will buy a spotty-looking apple because it is organic, and people who want organic but won't compromise on looks. Individual preference trade-offs can be more powerful in understanding what people actually do than the anonymous attitude studies or demographic segments associated with mass marketing.

Third, you want to understand actual *customer behavior.* It is important to get beyond verbal behavior—beyond what people say in answer to a questionnaire—to what they actually do. Past behavior is a better predictor of future behavior than almost anything else is. If you create micro-segments based on a database of actual behaviors, then you can look at the behaviors of new customers and make predictions about how they will act.

Thinking about "relationship"

To understand relationships in marketing, think about relationships between people.[5]

A relationship has an *ongoing commitment,* some kind of effort from both sides. When you buy a coke from a vending machine, that's a transaction. When you hire a lawyer, that's a relationship. The lawyer becomes *your* lawyer. The relationship outlasts the event that started it. You invest effort to help your lawyer succeed in working for you. If what the customer wants is just a transaction, there is not an opportunity for relationship marketing.

A relationship is about *mutual benefit.* If only one party benefits, the relationship dies out. Spaces where a relationship doesn't exist display lots of "churn." Customers move back and forth from one supplier to another, in response to promotions. In the early days of the cell phone business, before firms learned to pursue relationships, churn was as high as 30 percent per year. Businesses could not recover their customer acquisition costs before the customer moved on. Survival required some form of relationship.

Now think about the benefits of a relationship to the customer:

- A reliable source that doesn't have to be investigated before each transaction.
- Risk reduction, from someone he can trust to act in his own interest.
- Recognition, and perhaps a feeling of being valued.
- Access to information, education and service assistance.
- Personalized products or processes
- There may also be status or other emotional rewards from being associated with a prestige brand.[6]

The relationship marketer must keep finding benefits to deliver if the relationship is to continue—even if they are added-on benefits like those in a loyalty-card scheme.

A relationship requires some *knowledge* about the other party. It requires *interactivity*, and is stronger if there is lots of interactivity. Relationships are *responsive*: Feedback leads to behavior change. Because there is interactivity and responsiveness, there is learning. Relationships are dynamic, evolving. Their strength and character can sometimes be predicted, but never taken for granted.

A central issue in relationships is *fairness*. If the relationship is to continue, both parties need to believe that the benefits to each are in balance. Businesses that pursue long-term relationships with clients are not usually trying to maximize profits; they are trying to *optimize profits*. When you start making too much from a client, you look for ways to provide more services, to keep from endangering the relationship.

Another central issue is *trust*. In a relationship, each party gives up some control to the other. Customers must believe that a firm will act in their interests. A customer will share information only if she thinks it will be used to her advantage.

Frederick Newell says people judge a relationship with a business by two criteria: How much value does the relationship have for me, and how much control do I have over it?[7]

Businesses seem to pay more attention to offering value than to offering control. But in fact, control is a huge benefit as well. One basic way of keeping control in the hands of the customer is to give the customer access to the data collected about him. It will be argued that this is expensive and dangerous. But, in a digital world, customer information can be made available online nearly automatically to the person who is willing to go through the hassle of acquiring an identifier. Transparency to the customer is a powerful way to discipline the information collector, and keep the customer in control.

Businesses seem to focus on rational price-incentives more than on the often more powerful emotional rewards. Recognition of a customer's special status may be more important than the benefits associated with that special status.

Jim Barnes, in *Secrets of Customer Relationship Marketing*,[8] offers four measures of the degree to which a firm has committed to relationship marketing:

- How well can you identify your end users? If you can do this, you have the data to understand individual behaviors.

- Can you differentiate based on values and needs?

- Can you interact in a human way?

- How much can you customize?

Empower customer relationships with RFID

RFID technology can empower a relationship—rewarding the customer by making the experience better—and by doing so, create a reason to expand the relationship. It can enable smarter interactions, more personalized interactions, interactions in new places, interactions that offer new benefits and, perhaps most important, it can deliver all of these in real time—at the moment of maximum impact.

Imagine, for example, that you could persuade some of your best customers to carry a frequent buyer card, which bears an RFID tag, identifying them automatically when they enter your premises. This might sound intrusive. But it's happening at retail stores in India and Japan.[9] It can turn an anonymous customer into a relationship customer. And it can address each critical issue in relationship marketing

Customer acquisition works better with a personalized offer. Touch your RFID loyalty card to the kiosk at the supermarket and get coupons based on your own purchase history. Expensive shampoo brands target the purchasers of other expensive shampoo brands, and so on.

Retailers could do on-the-spot loyalty bonuses to aid *customer retention*. (See the Harrah's example.)

Water parks use RFID for *customer development*. They get extra sales with wristband-RFID payment systems that let kids buy refreshments without having to go find their parents. In the department store, a try-on mirror reads the customer's loyalty card and the tags on the garments she's holding, and lets her know about items that might go with them—just one or two—at a special price just for her.

RFID can change your offering based on *customer value*. Tesco supermarkets do valet parking for the customers they really need to keep. The loyalty card identifies the VIP.

You can redirect relationship marketing to fit *customer preferences*, as never before. Imagine an on-cart shopping assistant that identifies the customer who is particularly focused on environmental issues, and then offers comparisons of competing products based on an earth-friendly score.

Supermarkets have learned that a small chunk of the population does all the seafood shopping. Knowing actual *customer behavior* would let them offer coupons only to those who value them. And a real-time, in-store system could use those coupons to prevent losses from spoilage in this highly perishable category.

In fact, most of the dimensions of a relationship can be addressed. Amazon quietly demonstrates its customer *knowledge* by recommending books you might like. Border's could do the same thing in-store with its kiosk and the customer's RFID card. Home centers have contractor customers who are big-volume buyers, fiercely focused on price and on certain services. Individualized reductions, picked up at a kiosk with an RFID card, would improve the perception of *fairness*. The better the customer, the better the deal.

Some of these might be the wrong tactics. All of these might be the wrong tactics. But the point remains. It is possible, with RFID, to respond to the individual customer in a human way, with a customized offering. That's a giant step beyond blind, faceless marketing, and a giant step toward a relationship which empowers both customer and marketer.

CHAPTER 3

Customer Relationship Management works with RFID

When relationship marketing becomes a long-term strategic approach, it grows into customer relationship management (CRM). There are as many definitions of customer relationship management as there are practitioners, but most involve a combination of these five elements:

CRM is a collection of *policies, processes and technologies—*

to create and maintain relationships…

with current and prospective customers and partners…

across marketing, sales and service…

to optimize customer loyalty, efficiency, and the perceived value of the firm.[1]

In a sentence, it is the data-driven optimization of relationships.[2] Over time, it becomes more a point-of-view or an operating style than a formally defined process. But CRM does not belong everywhere. You have to recognize and stay out of situations where the customer just wants a transaction, not a relationship. Firms that can make the most of CRM generally have high-value products, products on which a customer depends. They have the ability to customize a product to some degree. They have the potential to earn incremental revenue from a current customer. They usually have high purchase frequency. And they can identify and collect information about a customer. That's not every business.

Note the differences from brand marketing. Marketing, sales, and service are tied together in CRM. Classical brand marketers, having fought their way free of the sales department, generally want to stay separate. Most would be willing to critique, but not to manage, the service function. And most see marketing communication as more important than sales or service.

The focus is on a relationship with the firm, more than on the product or brand. It is unlikely that Allstate Insurance would adopt CRM strategies for car insurance, but not for home insurance, though some procedures would be different. And you can have a two-way relationship with a firm, which you cannot have with a brand.

An emphasis on loyalty is different from, *and conflicts with* an emphasis on attracting new customers to grow market share.

Note also the reference to technologies. CRM is not just about software—too much attention to software was the biggest source of early failures in CRM. But CRM assumes that the power of technology will be there to make it possible for organizations to deal with large numbers of customers as individuals. The ongoing revolution that keeps cutting the cost of data processing is what made CRM possible.

Despite that, CRM is a high-cost activity. Smart CRM systems identify, prioritize, and build only the most important applications.

Principles of CRM and how RFID enables them

CRM has six basic principles. Each is transformed and empowered by RFID technology. Here's an outline, with RFID examples:

1. Identify and respond to the differences between customers.

Using previous individual purchase behavior, you can know who is profitable and who is not. With predictive modeling and maybe some data from secondary sources, you can distinguish between customers who are unprofitable now, and customers who are likely to be unprofitable always.

Using previous purchase behavior you can model differences in *preferences* among different customers. Here's what you'd like to know. Which preferences influence behavior? What are the customer's choices in the preferences that influence behavior? Which two preferences conflict, what trade-off is preferred? Which groups of customers are created by common preferences?

Using previous purchase and channel behavior, you can also detect specific behaviors that are hard to summarize as preferences: which products show up in every shopping trip, which promotions generate a response from this customer, and so on.

Much of this information gets collected via loyalty cards. They measure the frequency and recency of visits, the amounts and categories of purchases, and the success of specific promotional offers. They identify lapsed and declining users. They permit you to infer differences in needs or values between different customers. Mining the data collected by loyalty cards will reveal important but otherwise invisible differences that can be the basis for personalizing processes and building loyalty. With an RFID-enabled loyalty card, a brick-and-mortar retailer can know the kinds of things that e-commerce websites already know.

But only with RFID can you recognize customer differences fast enough to respond to them at retail.

Supermarkets hand the customer behavior-driven coupons on her way out. Not good. If she carried an RFID card, they could give her personalized coupons on her way in. Some people would get gourmet food coupons; others would get organic produce coupons. New moms would come in knowing they'd probably get a diaper deal. Coupons are often too small to matter. But if they could be targeted precisely, they could get bigger.

Another group of differences derives from how a customer was acquired. Customers acquired by referral are more valuable than those acquired by promotion. Referral customers spend more and value the brand more.[3] Suppose a customer could request a coupon to give to a friend. A brand marketer could afford to make an excellent offer on a coupon that travels with a customer's recommendation.

Customers acquired through a promotional offer are less loyal. If a fast-moving consumer goods marketer could know that a particular customer just gave his brand a try on promotion, the next few times that customer came in, the marketer would offer a bounceback coupon, to try to create a regular user. He could afford a rich offer to someone that he knows has already tried the product.

Long-term loyal customers are different than new customers. They do more of their buying at full price. They buy across the supplier's whole line; they cherry-pick less.[4] They're the ones who ought to get a special offer on the *line extensions* of a brand they've been using for years.

The key decision in relationship marketing is who to invest in. If you spend the same amount on each customer, *you cannot spend enough to create a relationship.* Every business has customers who will always be loss makers, and who should not be pursued. Businesses that are stuck in blind mass marketing will pay too much to acquire customers who will never pay them back.

Smart companies find an automated or near-automated self-service path for small or transactional customers, and offer more elaborate personal service to large accounts. Self-service technology is the key to cost-effective interaction. Think ATMs, kiosks, websites, cell phones—and RFID. NCR has a vending machine for movies on DVD. Tags on the disks make sure you get the right one and drive the re-stocking process. The absence of a salesperson keeps the cost down.

RFID has the power to change loyalty programs in several different ways. It can change the feeling of a relationship. It can move a customer from feeling like a promotional target, to feeling like a known individual. Combined with an access card, like a mass transit ticket, it can install loyalty-card marketing in ultra high-speed transactions, where there was no room for it before. And it can add the time-locating, place-locating function to enable instant offers that show up in front of the right person at the right moment to make a sale. It can also carry a behavioral history and relationship status without carrying personal ID.[5]

The UK's Boots the Chemist has a famous Boots Advantage Card, a smart card/credit card/loyalty card. The combination of a credit card and a loyalty card is not technically difficult. It can simplify the "spending" of loyalty points. It can facilitate special pricing for special customers. It can simplify the process of offering personalized promotions. And it gives you one less piece of plastic to pick through at the point of purchase.

With an RFID loyalty card, you can use previous purchases to guide promotional offers. Promotions that offer something extra with a purchase have an advantage over mere price cuts, because a fun or engaging premium can excite the customer's imagination, and provide an involving visual. But premium promotions are seen as risky because you have to pick a premium that appeals to your customer. When previous transactions guide that offer, you can get rid of the risk, but hang on to the extra engagement. Any loyalty card can make this happen, but an RFID loyalty card can work with events that happened today in this store, and in this very visit.

You can use previous purchases to create a preference profile. Hotels can know who wants a smoking room, who wants a room on a lower floor, who would not be pleased by a complimentary bottle of wine, who would love it if a free massage at the spa is thrown in. Tap your RFID loyalty card on the check-in kiosk and get a room that knows what you want, automatically.

Tesco, the supermarket chain, has a Club Card, which works with both perks and preferences. High-value customers get extras like valet parking. And deeply granular purchase data have differentiated 108 segments in terms of what they buy.[6] These segments can be used without personal identification to put the right offer in front of the right person at the right moment in a shopping trip.

In Mumbai, there's a start-up called Consumer Vision. It offers a tagged loyalty card to customers who are willing to have a store recognize them as they walk in the door. The tag can trigger alerts to specials and promotions tailored to that customer. Consumer Vision says the system works best in high-end retail environments where sharing some data is seen as a small price to pay for individualized special offers. It can be hard to get the customer's attention for an in-store promotion. The range of possible offers is huge, and the time spent in-store is short, and a dismaying percentage of customers walk out without finding something to buy. But Consumer Vision can select a few most appropriate offers from an arsenal of thousands. It can send those offers to a screen as the customer passes, or to the customer's cell phone. It can keep from putting an offer in front of a customer when the store has run out of his size.[7]

TeraData has demonstrated a method for collecting purchase behavior profiles based on individual market baskets, without using customer-identifiable data. The data allows stores to assess the effectiveness of promotions, to predict the propensity of a buyer who buys product A to buy product B, and to predict from a series of market baskets, the likelihood of that customer's attrition. TeraData has developed behavioral segments based entirely on the patterns of products purchased, not using demographic or geographic segmentation. For the retail marketer, this is like lights coming on in a dark room. It can improve both the customer's experience and the store's performance.[8]

Which customers cost you more?

Francis Buttle points out that RFID might permit activity-based costing.[9] Activity-based costing is a form of accounting which attempts to precisely associate costs with changes in the things your business does. Companies use it if they are thinking about discontinuing an entire type of activity or launching a new one. With RFID tracking service interactions, you could close in on the specific costs of *each thing you do for each one of a large number of customers*. Of course, if you had this kind of information, you could know what changes it would take in the services you provide to make a borderline customer profitable. You could see whether particular kinds of customization would pay off. You could know a lot more about whom to invest in.

2. **Focus on keeping and growing customers more than on acquiring new ones.**

Acquisition is both expensive and risky. It costs five to ten times as much to acquire a new customer as it does to retain an existing customer for one more year. In many businesses you have to keep a customer for quite a while to recover the cost of acquiring him. Successful acquisition is not a sure thing. You might spend a lot of money and not get enough new customers, or it might be that the ones you acquire are not profitable, or that the ones you acquire are profitable but not loyal.

Given limited resources, the relationship marketer will first fund retention, and then the growth of existing customers by up-sell and cross-sell. Acquisition comes third. Mass marketers worry about finding customers for a given product. Relationship marketers want to find products for a given customer. How much of the total spending of this customer in this category can I capture?

This shifts the focus from making promises bigger to keeping promises better. Do you know how well you keep your promises—in the customer's eyes? Don't count on complaints to tell you where you need to do better. Many dissatisfied customers never complain, and those who do complain may already have decided to move on. Retention requires early warning systems that customers can use to tell you you've let them down.

There are clear behavioral symptoms in almost every business that forecast a relationship's end. Cutting back on purchases predicts defection. Reduced frequency predicts defection. Partial payment of an invoice predicts defection. Complaints predict defection. And so on.[10] If you see it coming, you can often change the outcome. There is software designed solely for this purpose. Cell phone marketers mine data to identify the patterns of customer behavior that precede defection, and use some kind of special offer to try to pre-empt the defection.

RFID lets you try to pre-empt defection at retail in real-time. This is important for two reasons. First, pre-emption is likely to be easier if it is done more quickly, before the somewhat dissatisfied customer has made a definite decision to go elsewhere. Second, personal intervention at retail might reduce or remove dissatisfaction. An offer made by a live person has a much stronger emotional component than an offer made by direct mail, and does not carry the inherent annoyance of an offer made by telemarketing.

Win-back is the extreme case: the rescue of a relationship that has been lost. A service-business rule of thumb is that one third of your customers are in the process of dumping you. Lost customers are negative ambassadors. When you lose them, you also lose information about prob-lems that could cost you other customers. And there are only so many highly profitable customers in a market space. If you lose a good one, you limit future growth. Successful win-back efforts require three things: the ability to identify customers who are on a path to defection and are too valuable to lose, the ability to create offers that are compelling enough to induce a particular customer to give you a second chance, and a way to make the offer quickly, before the departing customer starts a relationship with a competitor.[11]

RFID enables all three. You can identify a defection pattern in real time, maybe while the customer is still in the store. You can select the most compelling offer from a pre-selected group based on purchase behavior up to that moment. Most important, you can, in some cases, make a win-back offer quickly and personally, while the customer is still on the premises.

A side benefit: win-back identifies recurring causes of attrition that you would not detect otherwise. If you know the drivers of attrition, you can fix them.

3. RFID enhances the ability to make relationships grow.

At retail, a shopping cart that reads RFID tags gives you market-basket analysis in real time. This kind of information creates cross-selling oppor-tunities. Think cream cheese with bagels. Sense-and-respond applications offer additional products automatically. Pick up a showerhead at the home center, and you get a coupon for the washers or silicon tape you'll need to install it.

In financial services, RFID promotes information sharing across departments. This improves cross-sell. Royal Bank of Canada, Wells Fargo, and many others focus their marketing efforts on moving an existing cus-tomer from an entry-level product, like a checking account, to other, more profitable products. With each interaction, they get a clearer prediction of the user's future needs, and are better equipped to make the next sug-gestion. With a card-driven interaction, they can make a suggestion while the customer is in the bank or at the ATM.

Because RFID is an automated and inexpensive way to promote to existing users, an enterprise can afford to try lots of different ways to grow a relationship, and then lean on the ones that work best.

4. Personalization is a key incentive and a reward for relationships.

I remember what you like and what you don't like.[12] That is part of what it means to be in a relationship. Whenever a customer encounters a process or product that has been personalized, the relationship gets reaffirmed and a barrier to exit is created or enlarged. Whenever relationship rewards are associated with a customer's identity, the relationship gets reaffirmed and a barrier to exit is created or enlarged.

The emphasis on personalization dovetails with the emphasis on retention and development. Personalization is okay for attracting customers, but it's great for retaining customers, and it's much more practical in situations where you've already collected some customer information.

Transaction learning over time can improve personalization and strengthen the relationship by making the service fit better, and making relationship rewards fit the needs of the customer. The car renter who will pay extra for satellite radio is the one who will come back more often if you throw in satellite radio as a loyalty bonus. The more explicitly you tie identification and recognition to the rewards, the more you teach the customer how to make the most of the relationship.

Peppers and Rogers speak of growing personalization as a "learning relationship," in which the customer teaches the company. The company acquires a competitive advantage by knowing how best to serve the customer. For the customer, loyalty becomes more convenient and the cost of switching gets higher.[13]

A key advantage of personalization is simplification. You don't have to collect the same data at every interaction. If possible, you never collect the same data twice. Promotions are fewer and more tailored. Relationship destroyers, like the current torrent of credit card mailings, are avoided. Simplification also cuts costs for the business.

Sophisticated personalization projects try to offer the same level of customized response across all channels. This is difficult or impossible if you can identify customers online, but you can't identify those same customers in a store.

Personalization can cut the customer's costs. The options he does not want, he doesn't pay for. Personalization can give customers better information. Different people want very different things at the supermarket. It's not easy to personalize the store, but with auto ID and a kiosk, or a cart-based shopping assistant, it is easy to personalize a customer's coupons, and thus easy for him to find savings on items he wants.

Much of the information needed to personalize can be collected automatically, by associating purchase behavior with identity, or with a particular cart moving through the store. Amazon quickly found which personalization services the customer values and which are seen as intrusive. RFID extends this process to every brick-and-mortar retailer and service provider. All it takes is the combination of automatic identification and digital signage at the point of sale.

Language and culture personalization is powerful in creating a relationship. Spanish speakers should see point-of-sale materials in Spanish, and the offers should reflect basic learning about multicultural marketing. The products you put on sale to increase sales volume from African Americans are not the same as the products you put on sale for American Hispanics.

Two critical limitations determine how far a firm can progress with personalization. First, high value, high involvement products or services have a better chance of creating a relationship. The more value a company supplies, the more its customers are willing to provide data. Second, the customer must feel in control of the relationship. People will only trade information for personalization if they feel in charge. This need not be an obstacle. Car rental agencies, airlines, online merchants, and ATMs all manage to give the customer enough control so that they are able to collect the necessary data.

5. A relationship must be managed across marketing, sales, and customer service.

At each place customers interact with your business, relationships are either eroded or enhanced. Yet organizations create silos of control that prevent service people from telling marketers that the product doesn't live up to the commercials. Silos of ignorance prevent relationship managers from absorbing customer information that shows up in somebody else's part of the organizational chart. Walls around customer functions are incompatible with strong customer relationships. Everybody who interacts with the customer must know what the rest of the team knows.

If you spend millions to create a powerful brand, you dare not tolerate a sales or service experience that does not enhance that brand experience. A customer-focused organization cannot do without a unity of purpose across all three functions—all focused on satisfying the customer's expectations. When relationships fail, everybody sees, after the fact, that marketing, sales, and customer service were not sharing information.

Both sales and customer service are better equipped to *listen* than marketing is. The more you shift into relationship marketing, the more listening takes precedence over "communicating."

Customer service is more powerful than most forms of marketing. One bad service experience may nullify all your marketing investment for that particular customer. Many times, for example, a company's call center is the defining experience in a relationship. The customer who calls is uniquely vulnerable—dependent on you to keep the promises you have made. Anybody can say, "your business is important to us," but over and over it is the call center that defines *for the customer* how much the customer is valued. Companies with time-wasting, disrespectful, and near-abusive call centers might as well stop spending on marketing. Their investment is destroyed by systems that ask you to listen to six long-winded alternatives, none of which is what you're calling about.

The customer experiences your entire company, unaware of the organizational barriers to sharing information. Your customer will assume that all of your company knows what the rest of your company is doing with him. Your customer will step out of a relationship, and even away from a transaction if you treat him nicely at one contact point and badly at the next. Your advertising will not move the customer if your call center has treated him like a jerk.

It's hard to create a set of processes in which an important customer is important to everyone in the company, but RFID makes it easier. There is nothing more important to a key customer relationship than recognition at every interaction.

It's hard to put the necessary information about a relationship in front of everyone who must contribute to that relationship, but RFID can make it easier. When I can see the customer, I can see her information, automatically, instantly, on-screen at the point of interaction. From her service preferences to her shoe size, I have all the information I need to serve her better, and *only* the information I need to serve her better.

It's hard to send marketing messages to the individual customer that reflect today's relationship, satisfaction level, or customer experience, and not the data of last year or even last month, but RFID can make it easier. It's already possible to update the online part of a relationship almost immediately. It's already possible to update the call center part of a relationship, almost immediately. Now it's possible to update every retail part of the relationship, almost immediately. If the patterns of events suggest defection is looming, I can intervene now. If the patterns of events suggest that a specific service disappointment has happened, I can intervene now. If the patterns of events suggest that my customer has just become ripe for a renewal or an upgrade or a cross-sell, I can intervene now.

It's hard to see the company's own service processes the way the customer sees them. But RFID can measure the service process at each step, and match performance against goals. Were the things that this customer usually wants available? Did she leave without a purchase, or with a smaller-than-usual purchase? Were offers made that recognize her value and preferences? How long did she have to wait? A new "concierge hospital" in Houston promises to get a new patient into a room in 30 minutes, and visited in that room by the nurse three minutes after that.[14]

It's hard to find the patterns in service interactions that lead to known, but unsolved problems. But with the mass of real-time RFID data, sometimes you can find a path to process improvement. As airlines shift to RFID baggage tracking, they will not just prevent individual problems. Almost certainly, they will end up reinventing the parts of the process that were causing problems.

Real-time visibility across all departments—whole enterprise visibility —is the way to overcome silos in an organization. RFID makes it happen.

6. The most important measure of success is loyalty.

Many businesses keep close track of some measure of customer satisfaction, but in the real world, satisfaction does not correlate with re-purchase. Lots of people who say they are satisfied are already trying out the competition. You can predict a lack of repurchase from very strong *dissatisfaction*, but it is not always easy to get people to express dissatisfaction.

What is important is not satisfaction but loyalty.

A very small change in behavioral loyalty—longer retention or bigger share of customer—has a massive effect on revenue and profitability. In Newell's research, a 5 percent increase in the number of customers retained for an additional year roughly doubles the profits of a typical firm. And that 5 percent increase raises the value of the average customer by 25 to 100 percent, depending on the industry.[15]

But loyalty is hard to win and hard to keep.

Customers have overwhelming choice. Our cities have too many stores. Our stores have too much assortment. A torrent of new products stuffs the shelves. The online environment opens up additional choices from all over the world.

Customers get more information than ever before, partly because of easy access online. Partly because of all this information, customers are more sophisticated and less in need of brand reassurance than ever before.

Brands have lost some magic. It is difficult to surprise and delight anybody. Yesterday's miracle is this morning's entitlement. Worldwide price competition squeezes the ability to get paid for any advantages a maker or a merchant can produce.

Time pressures force many purchases into a compelled convenience. No matter what customers want, much of the time they must buy what is available quick and close.

Loyalty is both attitude and behavior. We can better measure the behavior part: the consistent repurchase of a product, but we know that the attitude part is important.

Deals can buy repurchase but deals do not produce loyalty. Special prices or special add-ons may create trial and introduce a customer to a potential relationship, but the customer acquired by a deal is the least loyal of all, and must quickly be engaged if you are to earn back even the cost of acquiring him.

So in order to create the behavior, we need *incentives that focus on attitude*. Here are a few, from a much longer list by Tony Cram.[16]

- You can offer your best customers advance information, unavailable to others.

- You can provide some kind of private access—earlier or later or faster or more prestigious than that available to others.

- You can offer a simpler system to long-time or frequent users—unavailable to others.

- You can provide additional personal services.

- You can formally identify the loyal, and offer them special pricing based on their relationship with you.

One thing these examples have in common is that *all of them are made easier by RFID*. Advance information could be picked up from a smart poster, for example, and downloaded to an RFID-reading cell phone. Special access can be delivered automatically with a door or automatic gate that opens in the presence of a key customer's VIP card. An RFID example of a simpler procedure for known customers is the express airline boarding process offered at Dubai and other airports to carefully pre-screened and particularly valuable frequent-flyer customers. The card opens a door to a shortened process in which ID is double-checked, but the process itself is quicker and friendlier. Special pricing for best customers can be accomplished at retail, in response to a VIP loyalty card.

The importance of trust

Trust is the enabler of loyalty. There is no loyalty without considerable trust in the relationship partner. Trust is a subjective feeling of confidence in the future behavior of a relationship partner. It is created over time by a series of demonstrations from the relationship partner. It is revocable. It cannot always be rebuilt. At its center is the customer's expectation about your motives in situations where there is a risk to the customer and the relationship must offset the risk. RFID empowers and simplifies the activities that build trust.

Trust is enhanced by open access to information. I should be able to see the information that a company has collected about me, both online and in-store, if the company is a retailer. If there is something I need to do to earn a discount or award, I should be able to see that in-store. A lot of the problems people have with being "invaded" by one-to-one marketing would go away if the customer could know what the marketer has collected.

Trust is enhanced by individualized contact. Imagine the upscale clothing store whose sales associate could tell you, "This would also go with the skirt and shoes you bought from us this Spring."

Trust requires consistent behavior across silos of service, sales, and marketing. You cannot be a VIP customer at the check-out counter, and a person of no importance at the service counter. Maybe the people at the service counter need to see your RFID loyalty card.

Again, trust is not entirely rational. Many of the cues on which trust is built are not objective actions, but signals that the customer's welfare is being considered. The marketer who is careful about always asking permission before collecting information which will be used to produce better service is doing this kind of trust building.

Some limitations of CRM

The idea of Customer Relationship Management once evoked massive enthusiasm from marketers and managers, and even more massive spending, but much of the money was wasted and much of the strategy didn't stick. Out of that debacle came some expensive and important learning.

If you don't have a total commitment from top management to break down information silos, you cannot practice CRM. CRM requires active participation by marketing, customer service, and sales. Marketing alone can't force it to happen.

Human beings have no great desire to be stalked, scrutinized, optimized, and pestered. Permission and information must be paid for with clear, compelling, and sometimes expensive benefits. In some cases, benefits that seem compelling to the marketer will be turned down by the customer. Prada built a wonderful suite of data-driven aids to make shopping quicker, more convenient, less strenuous and more often successful. It could show a customer what a garment looked like at a distance and up close, share the fashion excitement surrounding it, and make certain that she knew exactly what her choices were. Prada found out that its state-of-the-art systems were not highly valued by its customers. Prada found that women who are willing to pay $500 for a pair of pants don't want you to keep any record of how big their bottoms are. No compromise was possible.[17]

The benefit to the customer needs to be seen *by the customer* as fair, based on what the company presumably gains. In some cases, the customer's idea of fair will be much more than you can afford.

Technology is intimidating. Data collection is intimidating. The motives of the company, however sincerely focused on building a better relationship, will always be suspect. So it must be clear at every step that the customer is in control and being rewarded. This is possible, but it is not automatic.

Cram sums it up neatly: CRM falls short where it is technology-driven, company-centric, or intrudes on the customer.[18]

CHAPTER 4

CRM 2.0 is Customer Experience Management

Classical mass marketing assumes that the brand is created and differentiated by marketing. Relationship marketers see this as unrealistic. The customer believes his own experience more than he believes advertising,[1] and the brand is defined by that experience. Competitive advantage depends on a consistent and satisfying experience. Service problems redefine a brand. Delivery problems redefine a brand. Process problems redefine a brand. Each is at least as strong as marketing.

The quality of the interaction may be as important as the quality of the product

Ask people what's most important in earning their loyalty, and in industry after industry, high-quality interactions with service and sales people are right alongside the quality of the product or service. Sometimes the interaction scores a little higher—sometimes the product scores higher. Both are far more important than price.[2]

Companies routinely have a better opinion of the experience they provide than their customers do. A Bain study says 80 percent of companies believe they deliver a superior experience. But 8 percent of their customers believe they are getting a superior experience. Another study says, in the past year, 80 percent of Americans have stopped doing business with at least one company because of a bad experience.[3]

Companies routinely have different ideas about what's important than their customers do. Companies say they're in a relationship business 98 percent of the time. Customers say they want a relationship 43 percent of the time. Compared with the customer, the firm understates the value of a personalized response. Compared with the customer, the firm underestimates the value of rewards for loyalty. Compared with the customer, the firm puts a much higher value on innovation and a much lower value on getting the basics right. Asked about quick response, 33 percent of customers are satisfied. Asked if the company can "understand my circumstances without my having to explain," 19 percent are satisfied.[4] These are scary numbers.

RFID creates unique capabilities to manage the experience better

Managers manage mostly in the dark. If you could know, moment by moment, which customer you are dealing with and what's important to that customer, and whether that customer is one of the ones who is keeping you alive—you could do a better job.

If you could instantly bring to hand the things that a customer wants to buy or may want to choose among—you could do a better job.

If you could effortlessly locate any tool or asset you need to provide a service to that customer—you could do a better job.

You can only deal with that much data if you collect it, define it, and process it automatically, and if the costs of doing so are minimal. That describes RFID. Companies that use RFID tags and readers for inventory tracking may find that their RFID system collects thirty times as much data as all the rest of the activities of the corporation put together,[5] but it doesn't cost a lot of money. What you can measure is what you can manage. And you can measure many aspects of the customer experience with RFID.

To brainstorm the opportunities, think, once again, about three core capabilities.

You can identify a specific instance of a product or asset or customer, and instantly point to a field in a database that contains any information you wish to attach to that single thing.

You can know the location of that tagged thing at every moment that it is in range of a reader, and you can put readers in lots of places.

You can know, with sensors, the condition of that tagged person or thing on any of a hundred dimensions from the temperature (of a frozen fish) to the velocity (of a dropped box) to the flow rate (of a bartender's whiskey pour), to the blood glucose level (of a far-away patient).

With those capabilities and a fair amount of ingenuity, you can then manage, as never before, what happens to your customer. You can identify the failure points in a process and automate them. You can identify the points where your process varies too much and stabilize them. Quickly. Automatically. And at just about any remote location in your control.

It's a tactical power, but it might change your strategies, because it makes things possible that were not possible before.

It enables event-based marketing and event-based experience management, where business rules define an instant response to the presence or condition or actions of a person or product or tool.

It enables business models where, rather than charging for an object, you charge for the stream of benefits that comes from a relationship.

RFID creates a better understanding of what's actually happening to a customer

You can know what happens during a store visit, fast enough to do something about it.

You can get a real-time understanding of the *context* of an event. When a customer goes to the home center and buys products A, B, C and D, you can see the project he must be planning, and check to see if he also needs E and F. He wishes you would. It would save him a trip.

You can get a much more granular understanding of the differences between customers—not just what she spends, but what kind of offers are likely to succeed with her, and what new products you should call to her attention. Not just how much and how often she buys from you, but what behavior changes might signal that she's upset with you and thinking about leaving.

Tracking customers' progress through a store or service location can improve the store's layout and design.[6]

At Harrah's, players insert a loyalty card into a slot machine. Managers get from this a vision of the casino floor, in real time, by age, spending level, game preference, drink preference, and so on. Automatic data collection makes it affordable.[7]

RFID creates a new ability to know things instantly

For years, CRM theorists have been preaching that the next step is relationship interaction in real time. Now it's possible. Now the old "storekeeper relationship" based on recognizing the customer at retail becomes more than a metaphor. Did the customer who always buys a bottle of Knob Creek walk past a shelf where it was out-of-stock? In the store of the future, you could see him coming and prevent that from happening. Did your bank's biggest depositor just walk in the front door? More than half of a bank visit consists of waiting. Maybe someone should walk her in and sit her down and help her out. Has the first-class passenger on a boarding flight fallen asleep in the VIP lounge? Easier (and friendlier) to find him and wake him than to delay the flight to unload his luggage. Do it instantly with an RFID frequent flyer card.

Customer expectations have changed. The online experience has made it reasonable to assume that, if you want to find something or find out about something, you can do it instantly. If you cannot find what you are looking for on a retail website, in a very small number of clicks—you simply click away and look somewhere else. There is no room in this new world for the discount store associate who says, "It's not on the shelf. There might be one in the back room, but I can't find it right now." Already a customer can walk into a bookstore and ask the terminal whether it has a book on RFID and the customer experience. Today, it can tell him yes or no. Tomorrow, it will tell him which aisle and which shelf and put a blinking light underneath the book. That changes the relationship.

Retail sales and service processes are mostly about information exchange. RFID permits *instant information exchange.* The store downloads info from the chip in a customer's phone, and sends other information back to the customer. Receipts, payments, customer preferences, new points to his loyalty program, papers that acknowledge he's returned the car that he rented. No paper, no waiting.

The big time-eater in customer service is collecting information from the customer about what you have done for that customer in the past. RFID can make that nearly instant. The chip on your laptop or washing machine or furnace can tell, at the service counter or over the phone, when you bought it and whether it's under warranty—and remotely diagnose what's wrong with it.

Baggage handlers need instant information too. When a customer misses his connecting flight and they re-route him home, his luggage may have only a few minutes to get someone's attention and say, "I'm over here and I need to be over there, *right now.*" No problem.

RFID creates a new ability to know things at the point of action

It's not what a business knows that determines the quality of the customer experience. It's what a business knows at the point of action—what the customer-facing manager knows in time to do something about it. Before RFID, a brick-and-mortar retailer or service provider could have a wonderful customer database and no way to apply it. Once you can attach that information to the person walking in the door, everything changes.

Some supermarkets in Europe can recognize a Gold customer entering the store.[8]

Some have permission to contact a customer in-store with a real-time offer. In the hotel business, there are RFID loyalty-card applications designed to deliver individualized services for high-value customers based on loyalty-card data. The card notifies the hotel when a VIP customer has arrived. It can put his or her picture on a manager's cell phone.[9] In software jargon, this is "embedded analytics." In the presence of embedded analytics, instead of a traditional marketing campaign, what you have is a series of chances to interact with the customer.

A new ability to create sense-and-respond applications

The extreme form of knowledge at the point of action is sense-and-respond. There are lots of situations where a particular customer action could produce an automatic response, as quick as an automatic door opener, but different for different customers. It could be as simple as generating a coupon for conditioner when you put shampoo into the shopping cart. It could be as simple as an ATM offering information about a high-interest, short-term investment product as the machine accepts a large deposit into checking.

Or it could be more complex. Capital One has software that identifies the source of an incoming call, and checks the characteristics and purchase history of that customer. Using datamining to compare that customer with others, it predicts the needs that customer might have right now, and the products that customer might buy. Instantly, it routes the call to the person best qualified to sell that product or solve that problem.[10] In that application, the identifier is a telephone number. With an RFID chip in a cell phone or credit card, the identification could take place at a store or service provider. With real-time analytics, the customer gets your best shot at the appropriate offer or reminder every time. You optimize every interaction. You don't steer the person who bought a flat-screen TV from you last month to a new low price on flat-screens.

The next step is Complex Event Processing. The software for complex event processing accepts data from multiple sources, combining information from a database with real-time information from the retail floor, for example—and uses it to make inferences about real-world events. Business rules define responses to these events. The responses come up automatically as a suggestion by a shopping assistant, a coupon generated by a kiosk, a product suggested by a sales associate, or any response that might offer a savings, customize a response, or otherwise strengthen the bonds of a relationship. A customer puts a package of pasta in her shopping cart and the cart offers a savings on the ingredients in her personal favorite recipe for pasta sauce. Maybe that's too intrusive, but maybe it would make her feel like she is in a relationship.

RFID enhances the ability to collaborate with others

Supply chain management, the fundamental RFID application, is an example of improving the ability of different enterprises to cooperate with each other, or of businesses to co-operate with their customers. But it is only one example.

Systems that show a customer the data that suppliers have collected about him could allow the customer to make that data better. Systems that show the algorithm a company uses to act on customer information may actually improve the relationships. As borrowers have learned about credit scores, for example, they have sometimes been able to take steps to make themselves more attractive customers. RFID can make this kind of interaction easier with a customer identification system that operates as an automatic password. Installed in a cell phone, it could collect data on the spot, when a purchase is about to be made, which may be the only place where the customer will bother.

There are promotion-tracking applications where a national brand marketer gets alerted to a problem in a particular supermarket, and works with the store manager to fix it. Things like that would be hard to do without RFID.[11]

RFID creates new power for self-service

Self-service applications have enormous power to cut costs and improve the customer experience at the same time. ATMs are so much more convenient than standing at the teller's window that you can charge the customer for the right to serve himself. UPS created a huge cost reduction by enabling people to track their own packages, and at the same time, gave them a better sense of what was happening and a way to reduce worries. The replacement of travel agents by self-service websites took out as much as 30 percent of travel costs, and gave people a clearer idea of what their alternatives are. The list goes on and on.

Self-service is particularly important in relationship marketing because it can transform marginal customers into profitable customers, and it can provide a better experience for the growth customer.

Some new self-service applications are created because RFID makes it possible to associate customer ID data with product ID data in real time. The "killer app" is automatic check-out and payment, where the supermarket cart notes the products you've loaded, and a doorway reader bills them to your payment card, and passes you an instant receipt.

The "car club" model of car rental is basically substituting self-service for the heavily manned office and lot of the conventional car rental company. There are many self-service rental possibilities. Home centers rent propane tanks, steam cleaning equipment, and infrequently used power tools. These can be checked in and out without need for an employee, with a library-like RFID model, which associates data from a customer ID tag with data from tags on the rented product.

Retailers can automate and simplify the process of exchanging a defective product for a working one, using tags that show that both are instances of the same product, and associating the data with a customer ID tag.

As the population ages, drugstores are trying to find ways to sell products like wheelchairs, electric scooters and hospital beds. No room on the sales floor? A kiosk with pictures and RFID real-time inventory is on lots of drawing boards.

Alan Estevez of the Department of Defense talks about how the dialog changes, between forward Marine units in a deployment and their logistical hub, when supplies are tagged with RFID. Before RFID, the forward unit is basically saying, "Where's my stuff?" After RFID, the communication is, "My stuff is in this place. Here's how to get it to me." The stakeholder now has the power to take care of himself.[12]

There are lots of online price-comparison products, listing products for sale on the Internet. The next giant step would be an engine that includes other instances of the same product available at stores near the shopper. The customer can order online or pick-up in store. This could be accomplished with item-level RFID.

Limitations

The most critical limitations on the use of RFID in the customer experience are privacy issues: limits on the appropriate use of information about the customer. These are complex and shifting, but not unsolvable. The next chapter spells them out and shows some ways to respond.

Adam Greenfield[13] has suggested another group of limitations on any form of ubiquitous computing. You can't assume the customer is paying attention. If the application breaks down, it can't get in the way. (A broken RFID tollbooth needs to be in the open position if it fails.) You need to let people know when they are interacting with an RFID system. (Theft prevention systems might be an exception.) You can't do anything that might embarrass a customer. (This happens every once in a while in database marketing. It is *powerfully* self-regulating.) You can't do anything that makes a process take longer. You need to give people a way to opt-out, just like they do in promotions that ask you to send in a box top. And you must make utterly certain that your machine is never impolite. The people at Citibank, who studied the ATM experience in excruciating detail, and spent a fortune crafting the human personality of their ATM experience, understood how important is the courtesy of machines. Relationship marketers had better do likewise.

Finally, in any application where you collect data, *keep the customer in control*. When customers can see their own data, you can eliminate the perception of stalking. You can enlist the customer in correcting his preference information. You can focus your relationship investment on people who want a relationship, and you can sometimes prevent the embarrassing mistakes that happen when companies jump to conclusions from transaction data. You can make customer data viewable online, in paper statements, or via cell phone. People whose credit cards accumulate miles or points or credits get that positive message at the point where they have to pay the bill.

RFID technology offers several ways for the customer to turn *identification off*. Extra cost is involved, but what a wonderful way to emphasize that the customer is in control.

Keep renewing permission. The only people you want to be interacting with, are people who see your extra service as a plus that they have chosen. Asking keeps them in control. If they don't trust you, then you cannot have their data. If they even suspect that you will get more out of the relationship than they do, then you cannot have their data. Make no mistake about it.

CHAPTER 5

Personal Identification and Privacy

Everybody who collects data about customers, and everybody who notices their data being collected, worries about whether we're all being kept track of too much. The people who write about RFID as "spy chips" touch a nerve. It doesn't seem to occur to many people that in a world full of watchers, the emergence of RFID technology might actually *improve* the ability of individuals to exert some control over the collection and use of information about them. That's what's going to happen, both in human identification and in the protection of consumer privacy.

The human identification dilemma

Lots of human interactions require the positive identification of a person not personally known to the other in the interaction. To pay a bill, cross a border, board a plane, drive a car, catch a fish, check out a book, write a prescription, own a gun, open a business, or enter a school, people need a way to show who they are that is acceptable to some other person. People have to identify themselves to others almost every day. They can't outgrow it or move away from it or stop doing it, whether they want to or not.

Identification is the process of getting someone else to recognize your identity. Usually it involves some *verifier* comparing your *identifiers* with previously recorded identifiers.[1]

We can't get along without it

Identification ensures accountability. There are lots of things you can't do or won't do because people know or can find out who you are. When identification is joined with recordkeeping, accountability becomes more powerful. Others can know not only who a person is, but also some slice of his actions in the past.

Identification plus recordkeeping create a way to manage risk. If I can find out that a person has lots of arrests for reckless driving, I may avoid renting him a car. If I can find out that he has declared bankruptcy six times, I may avoid issuing him a credit card. If I can find out that he has had a quadruple by-pass, I might not write him a big life insurance policy. Without identification and recordkeeping, car rentals, credit cards, life insurance and lots of other products would be more expensive for everybody. Some products could not be offered at all.

Identification permits the enforcement of laws and regulations. It is identification that makes it inconvenient for children to buy cigarettes and minors to buy alcohol.

Relationships between an individual and an institution require identifiers. The hospital needs to know which baby belongs to which mother. It goes on from there.

Identification becomes more important with the rise of global markets, and of remote transactions. I don't know who my customer is. I can't even see him from here. We can do business *only* because he possesses some means of identification.

But we need to get along without it

People prize anonymity. Anonymity is a great source of freedom—of relief from social pressures. Anonymity permits the expression of dissenting, controversial, and eccentric ideas with reduced social consequences. If we cannot maintain a balance between identification and anonymity, we experience enormous needless stress.

Identification joined with recordkeeping may give outsiders too much control over the lives of others. All individually identifiable recordkeeping is surveillance.[2] As data collection and storage became a thousand times less expensive, the amount of recordkeeping soared. As search technologies became more efficient, the accessibility of all those records soared. There is little control over the power of people and companies to collect information about other people. Many are uncomfortable with this. Anonymity permits freedom from personal history, from a train of data about every previous action and transaction that others might use to make inferences about us.

Finally, there are cultural variables. In a global marketplace, some cultures have privacy issues that exceed our own. The Netherlands has abolished the national census as too intrusive. People who live in countries, which once had or currently have totalitarian governments, are often sensitive about sharing any personal information at all with anyone outside the immediate family. These are real needs and they conflict directly with the desire of others to identify us every minute.

So how do we deal with a situation where we absolutely must have lots of trustworthy identification, and we absolutely must have as much anonymity as is possible? One simple answer would be: *We need the strongest possible identifying credential, and we need the greatest possible amount of control of that credential by the individual who carries it.* What we have right now is almost exactly the opposite.

Weak identification tools and little personal control

The United States has two major national identification devices: the Social Security number and the driver's license. Each was launched for good reason. Both have backed us into stupid situations.

The Social Security number now has more than forty congressionally authorized and government-required uses. It is not mandatory to have a Social Security number, but you can't hold a job, open a bank account, buy insurance, own a home, get a credit card, leave the country, pay your taxes, or buy a car without one.

Every time you use it, you increase the risk of its theft. It is difficult for twenty-first century Americans to go for a month without giving someone a Social Security number. Often they must give it to someone they don't know, over the phone. Then it goes into a database that lots of other unknown people can access. Almost no one could credibly claim that he can keep that number secret. But, with only a Social Security number and a date of birth, and some publicly available information, one person could apply for a credit card in another person's name, and use it to buy things that the name on the card has to pay for. (The number of an existing credit card is almost as vulnerable.)

Businesses who record your credit card number or your Social Security number in a database seem to lose them every day or two—a million at a time. Their pro forma apologies for butchering the financial security of their customers barely make news any more.

Almost every state has regulations that require the presentation of a driver's license to receive essential services, and other regulations that punish people by revoking their drivers' licenses for acts that have nothing to do with driving. Some states will take away a driver's license for failing to pay a library fine.

In the U.S., a driver's license is a "high integrity" identifier—accepted almost everywhere. But the people who issue driver's licenses have no urgent obligation to verify the identity of the person who gets one. To get a driver's license in Chicago, you have to come up with three "low integrity" identifiers: a utility bill (which is easy to create), a birth certificate (which is easy to create), a Social Security card (which is extremely easy to create), or several equally simple alternates. Think about the logic of using low integrity ID to qualify for a high integrity ID.

The only thing that prevents the mass forgery of these low integrity identifiers is the fact that it's quite convenient to forge the license itself. A police expert estimates that 50 percent of U.S. high school students possess a fake ID at some point.[3] The step-by-step websites say "produced with assistance from former members of the (STATE NAME) DMV." Get one delivered to your door for $150 or less. Or you can buy a real one from a DMV employee. Harper says the going rate is $2,500 to $3,000, somewhat higher in California.[4]

When basic identity devices cannot be trusted, problems occur. Twelve million people in the U.S. pay significant fees every year, just to make sure that no one has created phony credentials in their names.[5]

Identity fraud takes four forms:

Skimming is stealing credit card information in order to make unauthorized purchases.

Hacking is creating unauthorized access to a computer or network in order to steal information.

Phishing is the process of sending people to a fraudulent website to capture personal information. (A side-effect of phishing is the devaluation of information. Email users have learned not to care when notified that they have just won a million dollars in a sweepstakes, or that there is an urgent request from their bank for updated information, or that the IRS is notifying them of a refund.)

Pharming is the interception of personal information that is being input to a legitimate website. All of them attempt to use your personal information to take out fraudulent loans or to charge things to your credit card or the like.

By law in the United States, firms that make loans to a particular person bear limited responsibility for making sure who they actually give the money to. They have formidable power to compel the borrower to pay back such loans, whether he got the money or not. Firms that issue credit cards bear limited responsibility for making certain who got them and authorized their use. They have considerable power to compel the borrower to repay. Firms that provide credit information to lenders and credit card companies bear almost no responsibility for providing accurate information about a person, or for notifying the individual of the information they send out about him. According to Harper, a large proportion of identity theft actually originates with the employees of these firms.[6]

Lenders generally have "zero liability for fraud" policies of one kind or another. But an individual is guilty until he proves himself innocent. And an identity once stolen can be used again and again while the true owner of the identity will have trouble accessing credit.

Theft is only the beginning. With the rise in data that can be associated with an individual identity comes a rise in the potential for data mining—the use of individual data and software to discover patterns in the behaviors of individuals and to predict the propensity of an individual to act in a particular way. Data mining is relatively harmless in marketing. I am willing for people to make data-driven predictions about whether or not it is worthwhile to send me a cake mix coupon. It's more of a problem when governments do it, because data mining algorithms can generate a lot of false positives. If a marketer gets a false positive, he sends out a copy of the Victoria's Secret catalog to someone who isn't a good prospect for shiny shorts. If a government gets a false positive, maybe they pull somebody off an airplane.

It doesn't take paranoia to imagine a situation in which someone makes a mistake with a person's identification, which falsely includes him in a group of which he does not want to be a member, or falsely excludes him from a group he wants to be part of. Opportunities to review the information collected about oneself are few, inconvenient, and hard to find.

What makes a good method of identification?

All the ways of identifying people involve trade-offs. But you have to start somewhere. Following are some criteria based on an in-depth discussion of this problem in Harper.[7] A good identifier should be *unique:* no two people have the same one. It should be *universal:* everybody has access to one. It should be *permanent*—or at least easy to renew.

It should be *private.* Identifiers of the future must address a privacy problem that identifiers of the past have failed to deal with. You can find out too much, too easily about someone else's credit card use. *Secure* is different from private. A good identifier should be hard to steal and difficult for another person to make use of without the owner's knowledge. Most identification devices are *storable* somehow. But file size and speed of access are important. For years the FBI maintained a library of fingerprints, but it took days to find out if a given print matched one in the files. Biometrics may be reproduced on templates, in order to be carried in a card or stored in a database. These may require large files in order to be unique. An identifier should be *easy to collect*. Social Security numbers are easy to collect. DNA is harder.

A device that is *sufficient* will securely identify an individual without need for anything else. This is a tough one. As soon as you make a single document the only thing necessary for lots of occasions, you make it worthwhile to fake one.

Inexpensive. This is the most powerful metric of identification systems. The credit card identification process is nightmarishly leaky, but it's inexpensive for the card provider. The costs of its flawed design are borne by the merchant and the consumer.

Finally, an identifier should be *appropriate to the situation*. You cannot make too much trouble about identification unless it's truly necessary, and it is the customer's perception of too much trouble that matters. Marketers would like to restrict cumbersome ID to situations where there are large potential consequences and a high likelihood of deception.

The idea of multi-factor authentication

Authentication is demonstrating that a credential used for identification is valid. People do this by showing another credential that agrees with the first. If you are asked to provide two or three credentials, that is called "strong authentication," or multifactor authentication. It is presumed[8] to lower the risk of dealing with the wrong person. Usually people who are seriously trying to identify ask for "something you know," "something you have," and "something you are." *Something you know* might be a password or PIN number. *Something you have* may be a driver's license or passport. *Something you are* could be a biometric—a fingerprint, the pattern in your iris, the veins in your hand, the geometry of your face, or many others. All three types of authenticators have disabling weaknesses. *Something you know* is the easiest to steal. *Something you have* can be counterfeited. *Something you are* is expensive to collect. But a *something you are*, and either of the other two, or both of the other two, is powerful and hard to fake.

Private IDs and multiple IDs

A passenger cannot get on an airplane without a *government-issued* ID. Yet governments are notoriously sloppy providers. Individuals deliver government services. They are rewarded for speed, and for reducing complaints from customers who may be voters. They are not penalized for errors or rewarded for making the process more rigorous. Could private contractors do a better job? There are private programs that have improved both security and customer service.

ClearCard is a pre-screening service, a "known traveler" program which collects an iris scan, a fingerprint, a passport and some other documents in advance, and combines them in a smart card. Read at an airport kiosk, it provides access to an "expedited" ClearLane. Passengers still go through x-ray and metal detectors, but they claim significant time savings from shorter lines.[9]

Credentica produces a cryptographically protected ID token which can be carried in a smart card. It cannot be forged or modified, stolen, or replayed by the verifier. It cannot be correlated with other tokens from the same individual. It can be used to authenticate *one dimension* of identity, like membership, or loyalty status, without revealing anything else about the individual who uses it, not even a name. It can be given a limited number of re-uses, like a ticket. It can be added to an existing smart card that has other functions.[10]

Private providers of identity credentials face competitive incentives for both service and rigor which may not affect government systems. Better still, they tend to offer narrowly defined identifications. Multiple credentials with narrow applications make life much harder for the counterfeiter.

How RFID improves human identification

A new system for identification ought to address the failings of the current systems. It should prevent fraud and theft. It should maximize the user's convenience, and reduce or eliminate standing in lines. It should cut administrative costs, for governments and for businesses which the government requires to identify people. *Most important by far, it should increase the power of the individual to control his or her identity—to be easily and accurately identifiable when that is desirable and anonymous when that is desirable.*

RFID can be a platform for serious improvement. An RFID-based credential can be read automatically. The distance at which it is read can be precisely controlled. It can store large amounts of data in a tiny space. It can be fiercely encrypted. Consider the possibilities.

One proposal for an RFID passport comes with an on/off switch. It can be read automatically, while still in the passenger's pocket, so that he walks through the passport control gate of an airport with little or no line-standing. But it can only be read if the passenger wants it to be read.[11] When not in use, it's in the "off" position. Then it can only be read if opened, and only the printed part can be read.

That's a small step. Here's a giant step. Create a high-quality biometric identifier: an iris scan or a face scan, or a hand-veins scan—they're all convenient for machine reading. Make a template of that biometric and put it on a contactless smart card. You associate code with that template in the database. *Then you destroy the original scan*, and keep only the template in the card. It is a smart card: machine readable at a read-range of a few centimeters. That makes it automatic but really tough to skim. It has the unexcelled uniqueness of a biometric, but it's easy to store and easy to share. And if that individual walks up to a machine reader, and it reads his iris, and it reads his card, and they match, that's *positive* ID. No doubt about it. It's something you have and something you are and you can put a PIN number on it if you like the belt-and-suspenders look.

This biometric template doesn't exist in the database. It isn't in the software. It doesn't travel through the network. *It only exists on your card.* That is greater protection for individual privacy than anyone in the developed world has today. It can be protected from tampering. It can support digital signatures, for online authentication. It can support email encryption using an individual's private key. It can be read using existing

Windows or Unix applications. It can support the generation of one-time passwords.[12]

An RFID chip in the U.S. passport is being introduced. Many believe that the State Department's system is not the final answer, but it's a big first step. In the meantime, it is now real-world possible to supply the individual with an RFID credential that addresses every one of the criteria listed above for a good system.

An RFID card with the accuracy of biometric identification could be made to be readable at a checkpoint as a car drives slowly past. Vehicles of special interest could be stopped for a live iris scan to match the card. This would revolutionize both the accuracy and the convenience of border control.

A record of its collection is easy to store. It is so hard to steal that RFID-based payment systems like the Mobile Speed Pass have been in use for years without difficulty, whereas anyone who wants someone else's Social Security number can have it for the price of a case of beer. Cheap beer. Tom Espiner's survey of cybercrime says the going rate, online, for a complete set of credit card information, including account number and verification number, linked to the cardholder's Social Security number, is $10.[13]

The biometric for an RFID credential would require some effort to collect. But because it is permanent, the ID doesn't have to be re-issued. If you want it renewable like a driver's license, you can just write new code to the card.

Most important of all, an RFID credential nearly wipes out the trade-off between how much protection a business or institution needs, and how much inconvenience it can afford to impose. It is both easier on the user and vastly more powerful than the alternatives.

The privacy wall: Where it is, where it isn't

Everybody worries about privacy. Everyone can imagine being embarrassed or interrupted or damaged or annoyed by personal information that has fallen into the hands of people who shouldn't get it. Two studies spell out the key concerns.

A CapGemini survey[14] identified the four most-mentioned hotspots:

- Information exchanged between two trusted parties, which is intercepted and used by a third party.

- Harassment by intrusive marketers who have collected personal information, and use it rudely.

- The ability to track people's personal actions and whereabouts through items they have purchased.

- The interception of health or medical data that reflect on life-style.

A 2005 study in Germany focused on the *feelings* created where privacy is threatened.[15] Respondents talked about a sense of "powerlessness" when third parties accessed their information. They were apprehensive about how much further the technologies that penetrate privacy might go.

This sets clear limits, where marketers and governments may not tres-pass.

What is snooping and what is research?

There are lots of situations in which the seller would like to know more about the buyer. He might want to know more in order to sell that buyer more of the same in the future. He might want to know where he should send catalogs or post cards or emails. He might want to call the attention of current customers to new promotions on the products those customers regularly buy. He might want to identify characteristics which his best cus-tomers have in common. He might be able to decide, from learning about current customers, how products and experiences should be modified to serve those customers better. Depending on how they are pursued, these wants may be legitimate.

When research makes companies more efficient at meeting individual needs, the economy benefits and the customer benefits. This is not a trivial benefit.

Understanding what people want contributes to their happiness in important ways. Every serious attempt at improving the customer experience starts with learning more about the customer. It cannot be done in the twenty-first century supermarket the way it was done in the nineteenth century general store.

The question is, when, if ever, and how, if ever, may personally identifiable information be collected and used.

Anonymity and behavioral targeting

A seller may be interested in only one small fragment of a buyer's identity. Is this person (for example) a good prospect for fresh salmon fillets at my supermarket? It is possible to know this from loyalty-card data. *It is possible to separate this information from all other information about that customer, including the customer's identity.* In the MediaCart application (described in Chapter 7), a shopper shows his loyalty card to the cart. When the cart gets near the fresh seafood, it calls attention to that sale on salmon. *The system doesn't know who is pushing the cart.* All it knows is, this customer buys a lot of fresh salmon. The seller gets to target the behavior that matters to him. The buyer gets anonymity. There are lots of ways to make this happen.

Many times relationship marketers create an additional ID: a frequent flyer number or something like that. This is a powerful information suppressant. You can keep that number from being connected with other information about that individual.

Attitudes about privacy may be segmented

A Harris Poll suggests significant segmentation in people's attitudes about privacy.[16] It identifies three groups. For "privacy fundamentalists," privacy issues are critical and prohibitions are absolute. There is an implied willingness to impose the group's privacy positions on society, whether or not others agree. In the study, this segment is about 26 percent of the population. A second segment has no interest in or information about privacy issues. This group is about 10 percent. The remaining 64 percent sees value in privacy, but is interested in *trade-offs*. For this group, some personal information may legitimately be collected, but the user must pay for it, with privileges or price reductions. There is evidence that this segment of "privacy pragmatists" is growing.[17]

This suggests that there is room for relationship marketers to try to build a learning relationship with customers who want one, if the marketer does not go beyond what the customer permits, and gives fair value in exchange.

Where are trade-offs possible?

It makes sense to examine the trade-offs that people have already made.

When an Internet user goes online, she surrenders herself to certain kinds of surveillance. Someone is tracking where she goes, which pages of a website she opens, and how long she spends on each. Someone is tracking which keywords she searched, and putting up advertising alongside her search results. Many people are willing to go online anyway.

A cell phone user knows someone is keeping records of who is called, how long the conversation was, and the number from which it originated. To some extent, the location is monitored. The risk of illegal eavesdropping is greater than with a landline.

Holders of any kind of loyalty card know someone is tracking their purchases in that category, and could make deductions about their whereabouts and lifestyle.

If a car has OnStar, someone knows where the driver goes, how fast he drives, and how long he stays. The state makes drivers hang an identifying number on the front of their cars. Drivers can be photographed running a red light, evading a toll, driving too fast, or parking in front of the wrong house. People in large numbers decide to have a car, a computer, a cell phone, or OnStar service or a frequent flyer card despite its effect on privacy.

Health and Safety Trade-Offs. These are not usually controversial, though the data recorded is extremely private. Wristbands that make sure the right baby goes home with the right mother have been accepted where they are offered. There are few complaints about tags in a pacemaker that spy on a patient's heart rate and rhythms and share that data with remote medical technicians, or about smart cards that carry profoundly confidential health information for emergency hospital admission. Blood glucose sensor-tags that replace daily pinpricks have been welcomed. Workers in hazardous refinery environments are willing to have their movements tracked moment by moment through the day in exchange for the hope of speedy evacuation in an accident.

Cost-Reduction Trade-offs. A supermarket preferred-customer card can sometimes save $60 on a $300 shopping trip. That's the combined effect of a bunch of promotions and a relationship building strategy that is only partly about data collection. But from my point of view, it's a big reward for letting them keep track of what I buy. And if I later get a coupon in the mail for a product I've been buying a lot of, I do not resent the intrusion.

Special Services Trade-offs. Metro supermarkets offer valet parking for top-tier customers who show their contactless card. Most people can tolerate that; it's cold outside. A casino can read a gambler's loyalty card and tell him instantly if they will extend extra credit. Maybe he'd be luckier if they didn't.

Customization Trade-offs. The rent-a-car people know a driver's insurance data and car preference. They don't have to ask about it and punch it in every time.

Here's the point. If I can accept or reject the trade-off, if I can control the trade-off, there are definitely some trade-offs I will make. You can have my data if the price is right.

How RFID changes the privacy situation

Let's be clear: RFID is different. RFID is a step beyond bar codes, for example, in its potential to affect individual privacy. RFID technology creates the ability to obtain personally identifiable information about individuals, which would be difficult or impossible to obtain without RFID. So RFID applications *must offer* specific technological and regulatory methods to protect individual privacy. Here's what they must guard against:

- A reader could read a tag on an individual's payments card, loyalty card, key card or ticket, or on an item purchased, at a distance, without the knowledge of the individual.

- The purchaser of a tagged item might not know it is tagged, or might not be able to remove the tag.

- If a tagged item is paid for with a credit card, data exists which could tie the item to an individual.

- Tags identify unique instances of a product. A tag in your shoe would identify it, not only as a brown oxford, but as your shoe that you bought at a particular time and place.

- Tags can be rewritten remotely. Someone could alter information on a tag that you carry, without telling you.

- Tags on an item could allow someone to discover that it is in your possession.

The technological solutions

Radio waves, bouncing off a tag on some product in a store, a car, a customer's house or wallet, seem like a wide-open, risky way to transfer personally identifiable information. However, those who design RFID systems believe that the problems, which have been raised about personal identification in an RFID environment have practical, technological solutions. Following is a summary of the features of an RFID system that can be used to protect individual privacy. Judge for yourself.

Tag Killing. Many believe that the ideal solution is disabling tags as a product is purchased. There are easy ways to do this. Tag killing can be built into the retail checkout system, even if checkout is automatic. A tag can be scrambled after it is read. It can be disabled after it is read. It can be removed after it is read. There are several tag products which transfer control over a tag, for example on a consumer appliance, to the customer's cell phone or computer, so that the tag can be used later for warranty or maintenance information, but can only be unlocked by the consumer.[18] A point-of-service device can test and certify to the consumer that the tag has been killed.

Tag Blocking. A blocker tag makes it look like all possible tags are present already, so you cannot read that tag or any which accompany it.[19] You can use it to prevent eavesdropping. Another blocking technique uses materials in a wallet that block a long-range card from being detected.

Read Range. RFID tags can be designed to support read ranges of just a few centimeters. A tag with less than an inch of read range would be a pretty nearsighted spychip. At that range, you could read a bar code.

Tag makers, especially in apparel, have invented ways to give a tag a longer read range in the warehouse and the sales floor, but a super-short read range after purchase. In one version, the customer tears off half of a perforated tag after purchase. This removes most of the antenna, which cuts the read range. The customer gets tag identification if she needs to take the product back to the store, but pushes off any possible eavesdropper.[20]

In another version, the tag can be folded to shrink its read range.[21] It works like the tear-off tag, but is reversible.

Passwords. The basic low-cost Gen2 tag offers password protection, with 16 million password options. Passwords are in a different code than the rest of the message, so it's hard to steal a password by eavesdropping. Also, there are locking devices, so that you cannot steal data from a locked tag even if you have the password.[22] Before passwords, a retailer's tag data could be read by a competitor who entered the store, unless it was encrypted. Not any more.

On-Off Switches. There are tags you can turn on and off. On one proposed E-Passport, the owner has to push a button to let the tag be read.[23] Much less expensive is a tag with what is called a "physically changeable bit." This is like an on-off switch that you turn with your fingernail. There are healthcare uses where it turns off personal ID, but leaves other information in place.[24]

Encryption. Some see this as the simplest defense. Put the data in code. This is easy in a high-powered smart card. It used to be impossible in the cheapest tags, which have very little processing power. However, recent advances in cryptography have produced powerful ciphers with very low processing requirements.[25]

Database Defenses. This method seems simpler still. The tag contains only a "license plate." The data that give significance to that number are in a database, encrypted and physically guarded by the manufacturer or the home office of the retailer. That database need not have continuous Internet access. A thief would have to want the data badly to break into the factory to get it. Any transmission of the data can be encrypted and require strong authentication from the receiver.

"Unlinkability" is a new and low-cost procedure that keeps one chunk of data from being associated with another, without access to the mother database. It means, among other things, that an outsider cannot tell whether two signals are from the same user.[26]

The nuclear weapon in this odd little battle is another form of anonymization. Beyond unlinkability and encryption, anonymization is a method for scrambling data so entirely that it bears no resemblance to anything with meaning, and yet maintains the necessary "pointers" so that it can be included in an analysis of some pre-set characteristics. You could be looking at random gibberish and still be able to use it as input to answer questions about how many people bought the six-ounce size in menthol, and whether the coupon drop was worthwhile.[27]

Security structures in smart cards. In contactless smart cards, strong privacy protection is a byproduct of technology that was built to guarantee secure payments. Mutual authentication is the first step. The card and the reader must authorize each other. All information over the air is encrypted. It is verified to prevent modification or substitution. Only the minimum necessary information is sent. A firewall can be used to protect other information. A randomized number joins the ID number to prevent association of the number with an individual. There is a system to ensure that the transmission was a conscious action of the user. There is protection (with a random number on every transaction) against a replay attack. The chip itself has layers of defense including built-in tamper resistance, circuitry to block an attack through differential power analysis, and so on. Additional protection via biometric or PIN can be added.[28]

Tech solutions appear to eliminate clandestine privacy attacks

"Rogue readers," outside the retail store, will not defeat a two-inch read range, encrypted data, and tag data that only identifies a cell in a distant database. Tags carried in loyalty cards can be blocked, password protected, or simply shut off. A thrown-away tag that carries the encrypted number of a cell in the retailer's database is not worth the attention of a dumpster diver. There is no published record of a hacker attack on an RFID system "in the wild," that is, outside of academic laboratories with a point to prove. In most cases, there is nothing to be gained from such an attack.[29]

The vulnerabilities worth addressing are those created by permission— by the customer who enters into a relationship with a retailer, manufacturer, government, or service provider, and carries a tag as a passport to special privileges. In this situation, a structure must be set up to enforce the rights of the customer, and create explicit agreement about any rights accorded to the marketer. Even if the customer has given explicit permission, the marketer must work within these rules.

What rights do customers have?

Simson Garfinkel at MIT developed an RFID Bill of Rights which is echoed in almost everything else written about the subject.[30] He says, consumers should have:

- **The right to know whether a product contains RFID tags.** EPCglobal, the RFID standards-setting body has adopted a graphic to be used on signage and package to point to the presence of RFID.

- **The right to have RFID tags removed or de-activated when they purchase products.** See tag-killing, above.

- **The right to use RFID-enabled services without RFID tags.** If you have a toll gate on the highway that uses RFID-paid tolls, you must also have a gate with cash tolls.

- **The right to access an RFID tag's stored data.** This is important and neglected and goes beyond RFID applications. Customers whose data has been collected by any means should have the right to see, and comment on, data collected about them by a marketer.

- **The right to know when, where, and why tags are being read.** This presumably refers to tags read during the purchase process, not to supply-chain applications. The RFID icon addresses this.

The most important RFID rights are those established in existing law. In a nutshell, it is illegal to use RFID data to do harm to individuals, because it is illegal to do harm to individuals no matter how you do it.

By law, any collection of customer data must be secure—protected from unauthorized access, with a formal process for risk management, periodic assessment of new risks, and for the detection, reporting, and remedy of any security problems.[31]

Customers have the right to decide for themselves whether or not they will share data with marketers, in exchange for rewards. The supermarket loyalty card is an early but widespread example. The response to this from some professional privacy advocates has been that customers cannot be permitted to choose, that they are incompetent to protect themselves and must be forbidden to do so.[32] But forbidding individuals to dispose of their own data, even for what may be small advantages in convenience or small rewards in price, seems like a dangerous idea. Privacy advocates have the right to disagree with a customer's choices. But *they may not compel him to refuse a data-driven relationship* with any marketer of legal goods or services.

What rights do marketers have?

Professional privacy advocates, who oppose the very concept of automatic ID in the marketplace, speak in vague terms of the dark deeds that marketers will perpetrate with RFID data. Lots of colorful language about "voracious companies"[33] who run about the world "stalking"[34] the consumer and "capturing" her data gets sprayed around. It is generally modified by "have the power to,"[35] because real-world examples are not to be had. So it makes sense to be specific about the powers that relationship marketers believe they have.

The marketer has the legal right and obligation to keep accurate records of the amounts, dates, and terms of past transactions.

Other than that, *the only rights that marketers have are those that their customers grant them, often enough that they become customary. Any or all can be revoked.* By custom, then:

- Marketers have the right to keep whatever information they can collect about past transactions with their customers. Customers can pay with cash to preserve their anonymity. Health care-related transactions, prescriptions and the like, must be protected. The seller can study other transactions. That data may only be shared under explicit, published rules, which must be shown to the customer at frequent intervals.

- Marketers have the right to address communications to their customers. This can be limited, but only by request. Customers can put themselves on the "do not call" list for telemarketing. They can hire a service that forbids credit card companies and other lenders to send them applications that might be hijacked by identity thieves. They can throw away junk mail. But they probably cannot force marketers not to send it. Marketers are permitted to put up signs at the point of sale.

- Marketers have the right to study the needs and behaviors of customers to the extent that customers will permit. In the past, this was mostly sample-driven examination of "attitudes." Today, it is more often data mining, to discover or examine patterns based on previous behaviors.

- Marketers have the right to offer rewards to customers they identify as important. As customer data gets better, as marketers spend more to retain profitable customers than they spend on net-loss customers these will often be unequal rewards.

- Marketers have the right to address different messages to different customers. Habitual gourmets will see different information than shoppers with strong feelings about earth-friendly products, shoppers with certain nutritional issues, or shoppers with small children.

It sounds assertive to express these as "rights," and again, they may be removed at the will of the consumer. But in fact, this is all the "voracious companies" need to grow and retain relationships.

Who mis-uses RFID?

EZPass and the other automated toll collectors that use RFID tags to record trips through the tollbooth and send a monthly bill are hugely popular with commuters. They almost eliminate the wait at the tollbooth, merely requiring a slight slowdown. They don't make the driver come up with the right change. They save enormous amounts of expensive gasoline.

There is convincing research that says they have given consumers a more positive view of RFID technology in its other applications.[36]

That said, *this is a flawed and dangerous application.* It was started without sufficient forethought. It should be withdrawn and revised. RFID toll payments data has been subpoenaed in divorce hearings and disability hearings. It is data surveillance without adequate notice to the user. Unsecured tags in the existing system permit cars to be identified by rogue readers at non-tollbooth locations. Unsupervised vendor employees have access to personally identifiable data. Differential pricing for cash customers punishes the stranger and the poor.

Tags implanted in people should not be used for access control, as was done in one horrific instance by the drug-embattled Mexican government. They create an occasion for physical violence against the tagholder.

How can RFID improve individual privacy?

Drugstores are full of products that address embarrassing human frailties. Automatic checkout would be a mercy.

There are situations where tag-based monitoring could replace monitoring by an individual. One hospital application has shifted most of the monitoring of patients deemed at risk, away from security people to tags and readers. Tags and readers can't see when you scratch or burp or don't fix your hair and the labor savings paid for the system in less than one year.[37]

Systems that monitor the symptoms of seniors at home offer a huge improvement in privacy when they let a person spend more of his life outside of an institutional setting. This will be the biggest RFID story of the next decade.

The protection of personal financial data at banks and other financial institutions has become dramatically better with RFID. Vehicle tags, RFID badges, tags on bags of magnetic tapes, tags on computers that do not permit guarded assets to pass through doors without an approved carrier—these things have the power to prevent the giant and incessant data-security disasters of the last two decades.[38]

Identification cards that are RFID tags could be visually blank, keeping information from unauthorized eyes.

It's a hot niche. More privacy protection applications will appear in the future.

Best practices in a world of automatic ID

1. **Show the benefits to the customer.**

   Customer education is not optional. Privacy NGOs look for targets of opportunity. Attacks in the past have sometimes been one-sided, disingenuous, and non-factual.[39] It is a common assumption in public relations that advocacy organizations are more credible than the corporations they attack, so business users must get there first. A CapGemini study suggests that customers are open to RFID ideas, and opinions are not pre-formed.[40] Changing them after they are formed by someone else will be much harder.

2. **Customer notice.**

   Clear, visible notice that tags are being read must appear in every situation where information is collected about a customer or a purchase.

3. **Customer consent.**

   At every point where a customer could exercise choice, the customer must know about it. Where he can kill tags, where he can remove tags, where he can cut the read range, whether or not tags affect the warranty; wherever the data has some non-obvious purpose, the customer must be told. You cannot have RFID-collected customer data that the customer does not want you to have. Security applications and credit applications are the only exceptions.

4. **Customer access.**

   The customer must be able to see, quickly and conveniently, any data collected about her. Data disputed by a customer must be marked as disputed, and the customer's version placed alongside. Algorithms or scoring systems that offer benefits to the customer based on customer data should be visible alongside the data.

5. **Clear and stated purpose.**

   The besetting sin of database marketers is the collection of data that might be useful someday. As data becomes easier to get and the responses to data become more important to the customer, this sort of deep-sea trawling must be prohibited. If you don't have a reason, you can't have the data.

6. **Appropriate rewards.**

   You have to pay for what you take. Sometimes you pay with convenience, sometimes with customization, sometimes with a better experience. Sometimes you must pay with explicit discounts or extra services. If you don't pay the customer, you will eventually lose access to the data, and maybe the relationship as well.

7. **Security throughout the application.**

   Cryptography, mutual authentication, tamper-protection, read-range control—the mechanisms are in place for protection of the data your customer permits you to use. If you fail to make use of the tools at your command, the punishment is loss of relationship.

8. **Guard the data.**

   You may not record it if you cannot protect it. Customers have a right to know how you protect their data. As data-driven relationships grow more important, outside certifiers will emerge. They will build their business case on horrible examples. It is not a good thing to be a horrible example.

9. **Identify the data collector.**

   What scares people about RFID is when they cannot know the context in which their data is used. "Who" and "why" must always be available. "Who" is most important because it implies "why."

10. Explicit limits on sharing.

If you plan to share a customer's data with someone else, you must tell her who, how, and why. Specifically. This part of the process has been corrupted by useless, deceptive, and disingenuous "privacy policies" which are the practical jokes of corporate lawyers. If your privacy policy says, in effect, "we will only share your data with anybody we are working with (and we will work with anybody who will pay for it), and only for any reason we can come up with," then you are an exploiter and deserve neither data nor relationship. Many otherwise respectable businesses are dirty here. Fix yourself before you are fixed from without.

11. Explicit rules for sharers.

Being in a supply chain means organizing and monitoring behaviors among your partners as well as within your own company. It will be necessary to write contractual rules for the use of data from your customers, which are shared with your chain partners. You will need a mechanism to make sure your rules are followed. If your partners misuse customer data they obtained from you, you will be held responsible.

12. All personally identifiable data must expire.

It needs a use-by date. You have to justify how long you keep it. It should be blended into aggregated, anonymous data as soon as there is no clear way for it to benefit the individual customer.

13. A way to fix mistakes.

Before you start collecting personally identifiable data, you need to think about what mistakes you might make in its collection, storage, and use. You need to figure out what the customer would see as the appropriate response if data is disputed, lost, inappropriately shared, or used to drive responses that your customer sees as wrong.

The more powerful your sense-and-respond activities become, the better you have to be at preventing and fixing mistakes.

Many writers on CRM say that privacy protection should be a path to competitive advantage. Right now, most are playing defense, preventing relationship-destroyers rather than creating reasons to prefer. That's reason enough. It is no longer possible to have a relationship marketing strategy without a privacy strategy.

CHAPTER 6

Asset Tracking Creates New Business Models

COMPLEX COLLABORATION
The Special Case of Airline Baggage Tracking

You may not believe this, but airlines don't lose many checked bags. One study says 0.7 percent of bags go astray.[1] Others say it's a little over 1 percent.[2] Every year for four years, it's gotten worse. Today, a conservative estimate would be four million bags a year.[3] When it happens, it's almost as painful for the airline as it is for the customer. Directly attributable costs are about $100 dollars per lost bag. Sending a bag to a home or hotel the next day or late at night is expensive. That $100 does not count damage to the customer relationship, nor does it count security issues created by baggage separated from its owner. But $400 million a year is bad enough.

It is not a simple task to make sure everybody's bag gets sent to the right plane, loaded in time, transferred, and sent to the right baggage carrousel. Bar codes made a big improvement. But bar codes have some problems. Bags with bar codes need a lot more picking up and shifting around because bar code tags have to be twisted back and forth in front of a scanner. You cannot read a bar code on the bottom suitcase in a stack without moving the ones on top of it. You can't read a bar code if it is facing the wrong way. If there is a last-minute itinerary change, you can't change a bar code label without re-printing it. If you need to burrow into the belly of a 767 and find a particular piece of luggage, a bar code label will not help much.

Bar codes solve 80 to 90 percent of the problems that are likely to result in a missing bag. That's what got us down to $400 million. But RFID systems solve between 98 and 99 percent.[4]

Lots of airlines and airports are now testing and installing RFID baggage tracking. RFID tags don't have to be oriented. They can be read in the dark, and around corners and underneath other bags. They can be rewritten wirelessly to fit a changed itinerary or interline transfer. They let a baggage handler get one bag out of the hold in a hurry if needed.

One proposed system, TAG-IT from Texas Instruments adds a sliver of passenger control. Passengers could check to see if their luggage made it on the plane. They could make sure the routing was updated if their flight number or destination changed.[5]

If RFID can eliminate the problem of lost luggage, it will change the customer experience profoundly. If some early adopters can make a claim about lost luggage that competitors cannot match, they will have a powerful reason for preference.

Yes, you can buy a golf ball with an RFID tag and a handheld reader, so you can tell where it went when you hit it in the woods.[6] And that is only the beginning.

Businesses are just beginning to understand the impact of *asset tracking outside the supply chain*. When you can know, in real time, where things are, without sending somebody to find them, you can do things that you could not do before. You can get jobs done with fewer assets or less costly assets, or do things automatically that would otherwise require management on the spot. You can provide more satisfying service.

RFID makes it easy to know:

Where things are. If you need to retrieve one container from a freight yard, you can go right to it.

In real time. You don't need to record where things were, and then try to track them down. If someone in a hospital moves a mobile respirator from room 101 to room 909, you know where it is up to the moment it leaves 101, and you know when it enters 909. In some systems, you could "see" it from each room it passes on the way.

What their status or condition is. Sensors can travel along with tags, and report the status of a tagged item to a database from moment to moment, so you can know if your case of frozen fish was sitting in the sun instead of in the cooler, and exactly how long it sat there.

What other things are nearby. If you need to bring together a group of moveable tools on an assembly line, tags can show you the easiest combination. If you want to show the person who bought a size 46 sport coat where to find his shirt size, a retail display can do so instantly.

Other kinds of context surrounding the object. People who ship fresh flowers know that their shelf life may get shorter if the shipping temperature changes. An RFID tag with a temperature sensor attached can interact with a software application to figure out the change in shelf life, and re-route a shipment which got too warm to a nearer destination. This can keep a relationship from wilting.

There are different kinds of Real-Time Location Systems (RTLS). Some use a beacon tag that broadcasts a long-range signal, every few seconds. Some use triangulation; they send signals from interrogators in different places, and compare the time it takes for the different signals to get to the item and back. Some work alongside video systems, and can switch from radio to video as the tracked object gets close. Most asset tracking systems use long-range active tags. There is general-purpose software for asset tracking, and specialized systems for environments like hospitals or seaports. There are so many opportunities that RFID asset tracking has become a source of new processes and a breeder of new business models. The point of this chapter is to help you think about how such a system might solve problems for you or for your customers. The kinds of benefits described here overlap somewhat. They are organized the way they are to get you thinking about the ways RFID could improve your business model and help provide a higher level of customer satisfaction.

Some basic benefits

Management from a distance

If you have a tag and a reader on the spot, you might not need a manager on the spot. Consider a business based on renting things. You can track the comings and goings of a rental item in real time. Sometimes, that means you don't need a person standing there to check items in and out.

ZipCar is a car club—a shared service in which a car can be rented hourly. The car is parked on the street somewhere, at scattered locations across high-potential neighborhoods. The renter has a ZipCar membership card that contains an RFID tag. A phone call reserves the car and tells the renter its location. The customer's membership card gets remotely empowered to use the car. An RFID reader on the inside of the windshield reads the card and unlocks the door. At the door, the renter punches in a PIN number, which changes every time, to add another level of security. The renter gets the keys out of the glove compartment, and away she goes. Compare that with the old-style car rental company, which must lease a big facility at an expensive downtown location, with room enough to hold a lot of cars and people enough to service a lot of customers.

There are three new customer benefits: the ability to rent a car for just a couple of hours, the convenience of picking up a car right in the neighborhood, and the lower cost that comes from a more efficient operation.[7]

Airbus rents highly specialized maintenance tools to aircraft repair crews. RFID tags on the tools track their location, and sensors track their condition. They are important to maintaining the aircraft/airline customer relationship. But they are also a minute-by-minute revenue source. Their owner is an ocean away but because they are tagged, they don't get lost.[8]

A school kid comes home and touches his RFID card to a reader with a big sign that says "I'm home." If mom, at the office, wants to stop worrying about where he is and start worrying about whether he's doing his homework, she can query the reader remotely or have it call her.[9]

There are sheep stations in Australia where large numbers of animals are kept, with as little human intervention as possible. Each animal's tag tells its birth date, gender, ownership, and an identification number. The sheep have to cross a small bridge each day to get to water. The bridge is narrow enough to make the sheep pass through it one-by-one. Animals crossing the bridge are weighed automatically. If a sheep is not gaining weight, its tag opens a gate to a supplemental feeding station. Ultrasound devices focused on the sheep as they cross the bridge can see when ewes are pregnant, and whether they carry one lamb or twins. (Twins require assistance at birth). A facial recognition system checks for face worms, a common problem. Gates at the end of the bridge separate the sheep from the goats. Wild goats that have joined the herd get sent out a different door. The gates also separate out feral pigs, which eat lambs. All this can be accomplished automatically. Water provides the herding motivation. Clever arrangement of fences, readers and sensors does the rest. A visiting shepherd can address alerts generated by the system.[10]

Dr. Claus E. Heinrich, at SAP, predicts the rise of the "business envelope"[11]—his term for a process in which sensors track performance of a machine, engine, vehicle, or ship, measuring vibration, temperature, pressure, rpm, or the presence or absence of necessary fluids. Sensor readings feed an automated analytical process in a remote computer. The manager gets alerts when a process is nearing or exceeding its operating limits. It can predict the need for maintenance far enough in advance that maintenance can be scheduled efficiently. Preventive maintenance on a fixed schedule is not very efficient. Repair and maintenance during unscheduled downtime is horribly inefficient. Event-driven maintenance that identifies things that will have to be fixed in the near-but-manageable future can create a spectacular improvement. From a central location, you can monitor processes in places that are really hard to get at, down inside the engine of a ship, for example.

A German bakery puts tags on packages of ingredients and readers on the mixing kettles to make certain that each batch has the correct ingredients. If you leave out the cinnamon, it reminds you. But the manager doesn't have to be there to watch.[12]

Efficient sharing

If you know in real time where high-priced assets are, then lots of people, who may not communicate with each other, can share them.

Shipping containers are passed from hand to hand around the world. If you lose them, you have to make more. Identification numbers are painted on. People read these numbers, sometimes with field glasses, and record them manually. This creates errors. Databases tracking a 30-acre field full of containers stacked four or five high at a seaport or freight yard mis-locate or mis-identify about 30 percent of the containers.[13] An active RFID tag on each container can store its identity, dimensions, type, and weight. CHEP Corporation, which rents out a pool of pallets and containers has begun a massive deployment of RFID tags for its container fleet. Shippers rent containers by the day, and pass them along to the next customers.[14] The owner is not present.

Chemicals and gases are transported in high-quality rented contain-ers. Some gases are hazardous. Some must be handled in special ways. Bar codes on the outside of a tank wear off. Because bar codes cannot be changed, they may not be up-to-date on what's inside the tank. Finkenzeller describes applications where RFID tags in protective cases, fastened to the tank, carry information on the contents, ownership, maximum pressure, and so on.[15] The rented, reusable tank system is a practical way to cut costs. RFID tags provide the information to make it safer and more efficient.

Tulsa has a way to share bicycles, free for a day. Anyone can put a credit card in a reader to unlock a bike and use it for 24 hours. A tag on the bike and a reader on the bike rack show when you bring it back. There's only a credit card charge if you keep it more than a day.[16]

Mississippi Blood Services has a particularly tough set of problems to solve. Blood and plasma, donated for transfusion, expire in 5 to 42 days. Multiple types must go to multiple locations. Demand is completely unpredictable. It is possible (and, in the past, not uncommon), for product to expire unused even though it is currently needed.

Backward-looking batch inventories simply don't work fast enough in this environment, no matter how often they're updated. But RFID tags on each container of blood or plasma provide authoritative information on what is available *right now*. Sensors can flag temperature problems before they cause damage. The tag contains an FDA number, a product code, a unit number, expiration data, and blood type. It saves and delivers blood donations that would otherwise have been lost.[17]

Libraries are steadily adopting RFID despite enormous initial cost, because it eliminates most human monitoring. RFID lets patrons check out books without a librarian. It keeps books that weren't checked out from being removed. It makes re-shelving books quicker and easier. With a handheld reader, a librarian can quickly check a whole section and make certain that materials have not been mis-shelved by patrons or librarians.[18] Garfinkel reports a new problem: library patrons hide a DVD within the library so someone else won't check it out before they have a chance to. With RFID readers, you can quietly find them and put them back.[19] And librarians become what they want to be—people who help other people find things.

Process improvement

Often, when you see what is happening across a whole system in real time, you see a better way to organize the process.

One simple application involves a large auto dealership, which sets up cars on the lot so that they can be started with an RFID tag instead of a key. The salesperson can take a car and a customer for a test drive without going into the central building to get a key. The system tracks test drive activity by car model and by salesperson. If you know that a particular model gets test driven a lot, but doesn't get bought, you have an insight. Maybe it is mis-priced. Maybe there is an expectation that needs to be managed before the customer gets in the car. Maybe you need a better closing offer. If you know that a particular salesperson gets more test drives but closes fewer sales than others do, you know that certain skills need to be developed.[20]

The post office mails RFID tags around to see what really happens when letters are processed.[21] You can only manage what you can measure.

Construction has not kept pace with manufacturing in using data to increase productivity. RIFD applications enable serious gains. A construction site is a big place. You cannot simply look around and see what is there and what is not. Many tasks require a combination of assets: trucks and cranes and mixers and forms have to be in place at the same time, and tasks have to be accomplished in sequence. A few missing assets can cause a big delay and delays are no fun. A highway interchange out of service changes a lot of people's lives every hour until it's finished. Saving ten percent of the time over the life of a project can amount to millions of dollars. Tracking individual pieces of heavy equipment can change failure to success. RFID makes it automatic.

In the third world, it is common for heavy equipment to be in the hands of a particular driver throughout a project. Drivers earn extra income by slipping away from the site and helping out somewhere else. They give away equipment time in exchange for an under-the-table payment. Tracking equipment onsite makes the problem disappear.[22]

Insurance is a data-driven industry. Tagging assets and inventory will provide an amount of data at a level of accuracy vastly beyond what is available now. Students of the process believe that better data will provide a better picture of the real risk. This, in turn, will lower the cost of fire and theft and property damage insurance.[23]

Event-driven management

People who raise hogs say they all grow up at about the same speed. So it's smart to respond to any little variation. An RFID application directs tagged pigs across a scale as they are fed each day. The scale notices any pig that hasn't gained weight for two days. It sprays a spot of paint on the pig, and sends an email to the manager asking that this pig be visited. Maybe it isn't well. A first test of the system got 20 percent more pigs into the Hormel Red Box bonus category than without RFID. The value of reducing variation across the whole population was about 20 percent per pig.[24] Fat city!

A sensor reading could be the event to which managers respond. Consider the problem of the concrete contractor. Concrete is poured into forms and left to set until it is cured. Curing time varies widely, but if you could know the *moment* when the structure is cured deep down inside, you could probably cut some waiting time out of a construction schedule, saving hundreds of thousands of dollars per day. It is pretty easy to bury RFID sensors in the concrete when it is poured. Engineers have found a time-and-temperature algorithm that predicts how big a load the concrete can bear. Measuring the temperature inside the slab will sometimes let them re-open a busy stretch of highway a few days sooner, or cut the time between construction steps. That can change both cost and customer satisfaction.[25]

The big aquarium in Singapore tags fish. When a fish swims past, a screen lights up with information about that species.[26]

High-speed querying

Reading a tag is a quick way to collect information. It is especially powerful where the information you seek is hard to reach, or a huge amount must be collected, or when expensive processes have to wait while information is collected.

A tag on the top of a utility pole describes the equipment on the pole, its current condition and its maintenance history. Learning before climbing is faster.[27]

RFID tags are now being used to track buried cables. Atlanta's Hartsfield Airport has buried marker balls containing RFID tags every 200 feet along the route of a cable or pipeline—closer together near where the cables bend. Different frequencies identify different kinds of infrastructure. A chip can be buried as deep as five feet and still be reliably read. In a space like an airport, where redesign and reconstruction are expected from time to time, it is a huge advantage to have a simple way to know where to dig and where not to dig.[28]

Long-haul truckers make money by moving down the road, not by stopping and swapping data. But when they reach a terminal, they have to drop off a fair amount of data along with the load. One freight line has eliminated this long, unproductive pause with RFID tags on each truck and RFID readers on every terminal gate. As a long-haul driver enters the facility, the reader collects his data. A yard truck takes his trailer. He takes on a new load. The reader updates his tag. And he is immediately on his way again. Assets are used more productively and the driver collects for the extra miles completed.[29]

There are several proposals afloat to put RFID tags in currency—using a chip about the size of a grain of sand, with a very short read range. There are privacy worries. But in fact, currencies already have a unique identifier in the serial number. If it's worth someone's trouble to track $50 bills and associate them with people, it can be done with or without RFID. The RFID advantage is in speed and accuracy: counting large quantities of bills, and creating a record. It is not efficient to spend employee time counting money.

In the express package delivery business, the paradigm-changing innovation has been self-service tracking. Instead of providing a service team to track the progress of a customer's package, the shipping companies provide software that lets the customer track things online. This is more satisfying than asking where a package is and then deciding whether or not to believe the answer. And it is hugely less expensive for the shipping company. A next step in this process is being tested by DHL, which has a delivery van equipped with both RFID and GPS tracking systems. The GPS system shows online where the van is. The RFID system identifies an individual tagged package within the van. A reader on the door can send a message back to DHL when the package is removed from the van. The customer can look at a tracking screen online and see that a particular van is carrying the package, and where that van is right now.[30]

Pricing by usage

Knowing where an asset is opens up the possibility of selling a brief period of use. This doesn't have to be simply renting things by the hour. Norwich Union has a car insurance product that sets its fee based on an RFID measurement of how much you drive. If you don't drive much, they don't risk much, and you don't pay much.[31]

Alerts and reminders

Distinguished RFID nerds have amused themselves with home reminder systems that make a noise if a person is about to leave the house without his wallet, keys, and glasses. They used passive tags and a special watch. But the user has to remember to wear the watch.[32] It might be equally frustrating if he can remember all the things he is supposed to take with him, but can't figure out where he put them. It's not just the TV remote that falls between the couch cushions. Loc8tor tracks things for consumers. The user puts RFID tags on up to 24 things in his house. He gets a little reader to find them. The reader makes a beeping noise that gets louder when the reader is pointed in the right direction, and lines on the screen get closer together as the user moves closer to the object. Maybe this is silly, but two other competitors have already entered the space: Now You Can Findit! and iSpot.

Tags on poles along a highway send an alert if a car hits them. It gives time and location.[33]

Individualized products and services

Mass customization is the idea that products can be manufactured with a process that captures some of the efficiencies of mass production, but also with individual variations to fit the needs or desires of a particular person. Part of the task is keeping track of which variant belongs to whom. Firms which manufacture dentures, for example, can pick out the one for a particular customer with an RFID tag more quickly than with a paper label. The bigger the volume, the bigger the gain with RFID.

A giant hospital dispenses scrubs every day to many employees in many sizes. The employee's RFID badge opens a uniform storage closet. A reader notes who opened the door and which tagged scrubs have been removed. The reader keeps a continuous inventory within the closet and alerts the laundry team when more of a particular item is needed. Another reader at the dirty laundry bin takes custody of items from the employee. The hospital doesn't need a person to hand out scrubs, and it gets better inventory control than a person could have provided. Everybody gets the right size every time. Scrubs don't get carried off. It even takes less space than a conventional system.[34]

Personal records are an individualized product. If they can be found quickly when they're needed, high-priced professional labor can be used more efficiently. For years, businesses have studied the problems of filing records so that they can be quickly retrieved, but what about records that have been removed from a file? Whose office are they in now?

Advanced Pain Management is a clinical specialist organization with 28 locations. Its 50,000 patient records are stored centrally, but are often in use at a remote location. An RFID system lets you "see" where a record is immediately. There's an RFID tag on each patient's file, and a reader at the central records rack, on the door of the records room, and at each remote location. You can know if a file has left the room, and which office it went to. It's a simple system, implemented "out of the box," and it efficiently solves the problem of "who had it last."[35]

Maintaining safety

When you have to evacuate or rescue employees, knowing precisely where they are is life-or-death. But if you set up a check-in system that needs employee action to work, you can rarely count on compliance, even in the most hazardous environments.

British Petroleum uses UltraWideBand tagging of employees in refineries and on oil rigs to manage emergency evacuation. The employee has a credit card-sized tag in his wallet. He can't get into the workplace without it. The active tag, powered by a watch battery, emits one pulse per second for ten years. The UWB tag works well in a high-interference environment, which a refinery certainly is. Software superimposes the tag source on a map of the facility to show clearly the location of an employee who has not been evacuated.[36]

A new RFID application aims to improve the tracking of explosives in the supply chain. It uses both RFID and GPS technologies. The system provides a unique identifier for each detonator and explosive, plus another identifier for the case of explosives and yet another for the pallet of cases. It has biometric and RFID tagged identification for explosive delivery employees. It uses a GPS system to track the delivery truck. The truck is tracked in real time, and if explosives are removed from it, boom—you know exactly when and where it happened, and who was involved.[37]

Preventing losses and thefts

The RFID tag consumers carry every day is one they've probably never noticed. RFID tags in car keys prevent cars from being stolen. Chances are a car key with a black plastic head carries an RFID auto immobilization system. There's a tag in the key and a reader in the car, probably in the steering column. If the tag isn't there, the car won't start. It can't be "hot wired." An encryption system, combined with a rolling code procedure that writes a new pass-number into the key transponder (for next time) every time it is used, makes it hard to start a car without the owner's own key. The result is a rapid reduction in car thefts. Today, if cars are stolen, they may have been towed away, or they may have been started with the owner's key, but the traditional path of the car thief, to route around the key by touching some wires together, has been blocked.[38]

IBM puts an RFID chip on the motherboards of computers. Computers which are removed illegally are disabled. Information on their hard drives becomes difficult to access.[39] If stolen laptops cannot be used, maybe fewer laptops get stolen.

When you lose customer data, especially customer financial data, you generate a cascade of headaches. You have to inform every customer individually that his data has been lost, which is expensive and damaging. You may generate widespread media coverage, which will generally not be sympathetic, and may imply incompetence. There is a legal requirement to address foreseeable hazards. You may face legal liability which is difficult to quantify, difficult to limit, and difficult to defend against.

Despite a lot of attention paid to data theft through spyware, it is most often the physical media on which financial data is stored that are lost or stolen. Ten percent of all corporations, government agencies, and educational institutions suffered losses due to data theft last year. Eighty-three million records were stolen. In a single incident in 2005, a shipper lost backup data tapes with records of 3.9 million customers of a big international bank. Notification cost about $1 per customer, not counting damage to the brand.[40]

One service provider for banks uses a combination of asset control and asset tracking to safeguard backup tapes of financial data. Personnel, including delivery people, IT people, and administrative staff have RFID-tag badges. Vehicles have tags. Individual bags of tapes have tags. Data-bearing computers have tags. Doors have readers. If someone passes through a door without a current, valid badge, an alert is produced. If an asset is about to leave an authorized area, and no valid badge is present, an alert is generated and doors lock automatically.[41]

Other financial institutions are tracking records, checks, currency shipments, and bearer bonds.[42] Tapes that store back-up copies of data can be tracked by an RFID tag in the tape case, and the middleware can store a watch list of tapes not currently located, so that the system gets an alert when a reader sees one of them.[43]

Tamper-evident packaging has been a basic tool in loss prevention for more than a decade. Its tragic flaw is that someone has to *notice* that a package has been tampered with. RFID can fix this. When the package is opened, it instantly notifies the owner or shipper of the package. Knowing exactly when a package was opened will often mean knowing who opened it. Finding out about it the moment it happens may make it possible to apprehend the tamperer. And an easily visible alarm-stripe around a pallet will, in time, discourage potential tamperers.

Here's how it works. Plastic film is stretched over a pallet load of product. Printed on the film wrapped around the pallet is a metallic stripe. The stripe creates an electric circuit, with a circuit board and an RIFD tag attached at the end. The tag can be read from a distance, as long as the circuit is complete. The moment the film is cut, the tag stops being readable. There's instant notice of tampering. An alert gets generated. And there's a record of when it happened, even if the pallet is off in a boxcar somewhere. Current users include the Department of Defense, Wal-Mart, and many pharmaceutical companies.[44]

A drugs cabinet in an urban hospital contains tagged bottles of pills and medications. Whenever an item is removed from the cabinet, a reader triggers a digital photograph. The existence of the photo record is believed to change behavior.[45]

Making security more convenient

Designers of automatic ID systems believe that security should not cause delay, expense, inconvenience or fatigue. What we call security is mostly collecting information, comparing that information with a small set of known hazards, and preserving the information so that a terrorist or criminal can be prevented or apprehended. With RFID, you can do this without interrupting normal activities.

The enormous Port of Oakland admits trucks at nine different terminals—two thousand every day. The Port must collect the name and driver's license of each driver and the license number of each truck, coming and going. It must match the truck's number and the driver's ID with a permitted list in an existing database. The trick is to do this with a minimum of waiting. A voluntary program lets trucks buy an active tag that identifies truck and driver and can be read as the truck approaches the gate. This information is compared with the database, and the driver stops just for a moment to show his driver's license. The system reduces waiting time for everyone, whether they have a tag or not, but the trucking firms know that the more trucks with tags, the shorter the wait will be. Eventually, additional readers will be installed to track where trucks go within the port.[46]

Empowering collaboration

Sharing data can help even very large groups of people work together. The trick is standardizing the format of the data and the definition of terms, and making the data available to everyone instantly. This is easy to accomplish with RFID.

The Dutch Horticultural Coop is a giant auction exchange where producers sell flowers to distributors or retailers. Flowers arrive and are sent on within hours. Flowers are graded by an agreed-upon set of standards by exchange employees. The grading and other descriptive data goes on tags. Data is time-stamped to quantify freshness. The flowers enter the exchange on numbered trolleys. Antennas embedded in the floor note the arrival of a particular trolley. The trolley number is linked to the tag data on batches of flowers. Data about trolleys and their contents, from tags, are posted on large LED screens, in front of the auction buyers. There is no faster or more efficient way to put the data in front of the bidders.[47]

The giant, open marathons conducted in many U.S. cities also require a high level of collaboration. One of the problems is how to measure the starting time. An event that has thousands of entrants cannot line them all up at the starting line. But every runner wants to know his or her exact time. Another problem involves making certain that all runners who cross the finish line have completed the entire course. Both of these are good RFID applications. A tag on the runner's shoelace provides an exact start time and finish time, based on readers at each end point. It is not affected by large numbers of people passing the same point at the same moment. Readers along the course keep a record of when a tag has passed a given point. As an additional benefit, tags make it easy to post a leader board along the way. It would not be conceptually difficult to set up a website that lets a spectator track friends or relatives and get to the right point at the right time to cheer them on.[48]

Matching-up elements in complex systems

Everyone has learned to connect the right cable to the right port to get computer components to talk to each other. But imagine a system with fifteen or twenty thousand ports and cables, where ports get their names changed frequently, and are labeled only inside the computer. Imagine a system where there are damaging consequences if you accidentally touch a cable to the wrong port. Now you need tags and readers. An interrogator at the port with an extremely short read range can light up a cable-tag when it gets close, so you quickly find the right one. If a cable comes loose, it stops a radio signal and creates an alert that lets you know it came unplugged and exactly where it came unplugged.[49]

Next step: the home network. Maybe a user shouldn't have to plug in all those peripherals at all. The wireless router in his home office gets a reader, and every other piece of computer stuff in his house gets a tag. Proximity creates the connection. So he can print to any printer in the house. Tagged devices turn themselves on when they are accessed from the computer, and the rat's nest of wires under everybody's desk goes away forever.[50]

Hazard detection

Tags with sensors can generate an alert for a predictable dangerous event. Sensors can monitor tire pressure, detecting slow leaks that even a sharp-eyed and attentive driver would miss. A chip can be bonded to the inside of a tire. This makes it tough to tamper with.[51] Sensors carried by workers in nuclear power plants can detect radiation, as well as providing real-time location information for worker safety.[52]

Sensors on shipping containers can tell if the doors have been opened. They can provide an instant alert, and a record of exactly when the door was opened. They can also detect changes in temperature or humidity which might damage the contents of the container. The Container Security Initiative is a systematic proposal to use RFID tags to track shock, temperature, light, the presence of radiation and of particular chemicals, as a defense against terrorism.[53]

An RFID tag with a thermal sensor updates and networks the home smoke detector. Using a temperature sensor, it can notify firefighters even if nobody's home.[54]

An interesting new application in the management of casinos uses a biosensor on a wristband worn by employees. It gives an alert if the employee's heart is racing. This causes a camera to be turned toward the employee. A variant of this is a sort of silent alarm at convenience stores. If an employee is frightened enough to change his heart rate, it sends an alert to outside security.[55]

As RFID tags, sensors, and in some cases, readers, become part of cell phones, it may be practical to carry around a battery of sensors to address an individual situation or personal worries. Cell phones can carry a radiation sensor, a heart monitor, even a carbon monoxide detector. They can call the police if someone is entering a home.[56] There are so many healthcare applications that they need their own chapter.

Monitoring the cold chain

Basic to supply chains in foodservice, grocery and health care is the concept of the cold chain moving products from fishery, farm, or factory through warehouses and distribution centers to the customer at a restaurant, store, or hospital, while maintaining a constant temperature. Monitoring this process is harder than you might imagine. Mechanical devices for logging temperature data are expensive and difficult to maintain.[57] The temperature at one end of a shipping container or truck trailer can be much different than the temperature at the other end. Some kind of monitoring is critical. Foods and medicines can be damaged or destroyed if the cold chain is broken. Knowing how long a lapse has lasted is critical to knowing what to do about it.

ColdStream measures temperature in the trailer of a refrigerated truck every ten seconds. It can send an alert to the driver, and keep a record for the receiver.[58]

Chile ships avocados 6,000 miles by sea to the United States. Avocados are very sensitive to a breach of the cold chain. If they get too ripe too soon, grocery chains won't accept them. A ship can be a difficult environment in which to preserve the cold chain, and an untrustworthy place to monitor. Rio Blanco, one of the big exporters, has introduced an RFID application called "paltags." ("Palta" means avocado.) It uses individual case tags and handheld readers and a set of rules for continuous temperature checks. Paltags provide reassurance to the receiving grocery buyers and may create a brand preference at the wholesale level.[59]

Cutting the cost of watching product

A lot of the relationship problems that crop up between suppliers and shippers and distributors and retailers stem from the fact that it's hard to monitor products in motion.

Who dropped it? Who stole it? Who thawed it? Who opened it enroute? In the past, it has generally been too hard to know what really happened and too easy to solve the problem in favor of whoever has the most leverage in the supply chain relationship. As a result, flawed processes don't get fixed and suppliers who feel victimized by their downstream customers practice defensive pricing.

Real-time visibility shines a bright light on process problems. If you know when the problem happened, then you generally know who needs to change their ways. The great contribution of RFID to this process is to make it affordable to follow the condition of products from moment to moment. One of the easy ones is "Who broke it?" An acceleration sensor can tell you when it got dropped and from how high.

You can monitor conditions inside a crate on a high shelf in the remote vastness of a warehouse for not much money and probably fast enough to fix a problem before it causes damage. You can monitor the pressure of fluids in an enclosed container, or even in a pipeline, and know if you've sprung a leak.

It's hard to know how far this will go—but certainly much further than it has gone so far. Auburn University is developing sensors for food-borne bacteria. Bacteria are deadly, destructive and hard to detect. If a bacteria sensor can be fixed to a tag, it could have the impact of a new miracle drug.[60]

Sensors and the management of dynamic shelf life

The shelf life of consumer products has to be carefully managed to protect the customer, the retailer, and the producer. You don't ship flowers across the country if they're going to be almost dead when they get there. You protect your retailers from receiving a product just before it expires.

This fairly simple set of rules gets much more interesting when you can measure the condition of products in motion from moment to moment. For some products, it is possible to calculate how shelf life will be changed by conditions on the road. If a trailer full of fresh seafood loses its cold environment for a couple of hours, it is not ruined, but its shelf life is shortened. If a cargo of pharmaceuticals is held at an inappropriate temperature, its use-by date is changed. In both cases, there is a way to look at the conditions and calculate the shelf life changes. It may make sense to change a product's destination. Medicines might be re-routed from a low-volume destination to a place where they are likely to be used immediately.

One RFID label-maker produces a credit-card-sized tag that logs data on the intensity and duration of exposure to light, temperature, and humidity. When the tags are read, middleware adjusts the shelf life and notifies the sender. Trials are taking place with pharmaceutical products and perishable groceries.[61] It solves relationship problems for suppliers and retailers, and protects the customer from expired products that aren't labeled as such.

In a world where businesses focus more and more on their core competence and outsource everything else, in a country where security needs and convenience needs both demand the highest priority, in an economy that can only be competitive by substituting digits for labor wherever it is possible to do so, in a society where personal service by a physically present individual becomes more and more of a luxury, the need for automated asset management will continue to accelerate. RFID is the answer.

CHAPTER 7

RFID and the Retail Experience

MEDIACART®
Interactive POP

If you're going to sell a shopping cart that costs a thousand dollars, it had better do some pretty fancy shopping.

MediaCart thinks its new product might be a bargain at the price. They reason thus: 70 percent of purchase decisions take place at the point of sale. Brand marketers, trying to be where the action is, have curbed their spending on advertising, but doubled their spending on efforts at the point of sale. Now comes a product that can make life dramatically easier for the shopper, and can put the brand marketer right at her side—with a message, and maybe a coupon that pops on screen just before she reaches the product in question. The message is "behaviorally targeted"—aimed only at shoppers whose previous purchases make them look like the right people to talk to.

MediaCart is a computer on a shopping cart, with a 12-inch screen facing the shopper. The top half of the screen delivers shopper services. Customers can speak the name of a product and the screen will show them where they are right now, and where to find that product. They can check a product for calories or fat grams. Input food allergies and get an alert if they are about to buy something with an ingredient someone in their family is allergic to. It can keep a running track of what the customer has spent so far. Customers can check the price of a product against its competitors, instantly.

Customers can type up a shopping list on their home computers. Associate it with a customer's loyalty card or PIN number, scan the card or punch the PIN on the cart, and the list is arranged in aisle-by-aisle order, so she doesn't forget things and doesn't have to retrace her steps.

MediaCart can look at what customers have bought so far, and propose a recipe, in time to let them pick up any remaining ingredients.

Here's the biggest customer bonus. If the supermarket is willing to support automatic checkout, MediaCart can check them out in about 12 seconds—with fifteen items or a hundred and fifteen.

Now think about the bottom half of the MediaCart screen. It can show "banner ads" or video: seven-second or fifteen-second commercials. It can put them up at the right moment in the shopping trip, shortly before customers get to that product on the shelf. It can even throw in a coupon.

And here's the biggest advertiser bonus. It can direct that commercial only to people whose previous purchases show that they are good prospects. Not everybody buys super-premium shampoos and conditioners. Only if the customer is one of them would he or she get the commercial—and the coupon—introducing a brand new one.

MediaCart's CEO claims a 90 percent recall rate for those messages. Well, no wonder. The customers are right at the point where they make a buy/don't buy decision. And they didn't get the message in the first place, unless they were proven users of the category.

Messages and offers will be measured immediately. If they are not driving purchase, something can be changed.

And MediaCart can deliver that individually tailored message without invading privacy. If customers are regular buyers of high-priced cat food, it may send them a high-priced cat food coupon, based on the data from their loyalty cards. But the store won't know, and the brand won't know, and MediaCart won't know who the customers are. All they will know is that the offer went to a cart being pushed by someone who sometimes buys in this category. Customers can also choose to use MediaCart without sharing their loyalty cards, but they'll miss a few well-chosen coupons.

Right now, MediaCart works with a bar code reader. But it's RFID ready, as soon as enough grocery products are tagged.

It is relationship retailing made automatic. It tailors offers to customers, in-store, in real time.

A hundred and fifty years ago, when a customer walked into a retail store, she entered into a real-time relationship experience. There was little "self-service." She could expect a social interaction with a trained professional. Often as not, she dealt with someone who knew her, knew something about her values and brand choices, knew enough to remind her if she was about to forget something. The sales person (often the owner) knew the alternatives to a purchase, and the complementary items that could be sold at the same time. He knew the store's inventory well enough to anticipate most of the questions people asked. He knew what to offer if his customer wanted to spend a little less, or if she might be tempted to spend a little more. He knew what a particular item had cost the store, and had the power to drop the price a little if necessary, to satisfy a customer and close a sale.

F.W. Woolworth and the merchants who followed him eliminated what they called haggling. They switched the focus from relationship to efficiency, and it never came back.

Today even the most relationship-conscious retailers are designed around anonymity and self-service, maybe with a little help, diffidently offered, in a few exceptional categories. Customers are so used to this that any personal service at all may be startling, almost intrusive. Recognizing a specific customer is risky business, and it happens only if it has somehow been pre-approved by that customer.

But imagine what might happen if a store could earn the right to recognize a customer and serve that customer's individual needs.

Relationship retailing attempts to deliver an individualized experience that is customer-directed, context-aware, and identical at every point of contact. Obviously, this can be done only with customers who want such an experience. Relationship retailing is probably limited to VIP customers, though it could be broadened over time. Here is one perspective on how it might work.

The customer is identified as he or she enters the store, either actively, with a contactless card that the shopper waves at an entrance kiosk, or automatically, with a door reader that sees the RFID-tagged loyalty card of each relationship customer. Some retailers will start with a structure in which the customer identifies herself if she wants to, and shift the system to automatic identification of relationship customers after people are comfortable with the idea.

Information that the customer wishes to share, like shoe size or a preference for organic produce, is instantly available at a shopping-assistant computer, which is on the customer's cart or at a device carried by a sales associate in stores that offer personal assistance. This is not just an aggregation of old data, combined with data mining, but dynamic data that is updated in real time as the customer moves through the store. The data permits personalized alerts and suggestions. "We have three *other* red sling-back shoes with three-inch heels in size six. Here's a photo of each. Would you like to try any of them?"

Experienced users may permit more intrusive alerts. "Did you notice your preferred brand of cat food is on sale?" Experimentation will determine which alerts are seen as helpful and which are annoying. The shopper can touch a button to say, "Don't bother me." Or "Don't make that suggestion again." Shopper responses enter the database and influence future alerts.

In addition to personalized information, the system can be context-aware. It can know that it's back-to-school time for the customer's three kids. It can know that it's baby Tyneshia's birthday next week, or that last year, just before Easter, this customer bought several bags of candy eggs. A context-aware system with instant updating can offer *dynamic couponing*. A box of cake mix dropped into the cart produces an electronic coupon for a particular brand of frosting mix. This happens, not at checkout, but before she passes the baking supplies display, because part of context-awareness is being aware of where the customer is on the selling floor.

Here's a more controversial possibility: a context-aware system can know previous responses to promotions, and act accordingly. Some customers are coupon-prone; a small incentive will change their behavior. Other customers will respond only to a larger price reduction. It would be possible to tailor the size of the coupon offered based on previous responses. It would also be possible to tailor the offer based on long-term customer value. Should the merchant attempt either of these? Right now, there is little data but lots of opinion on this topic. Some have suggested that individualized price cuts, expressed as *loyalty bonuses*, will be a basic way of rewarding customers for sharing their personal information.

In relationship retailing, the same information is available at every channel, so if your customer wants to return an item that she ordered from the catalog to the retail store, that's not a problem. If she is a VIP customer entitled to special perks or discounts, everyone at every branch knows that. If any part of a transaction requires collecting data from the customer, it only happens once. She isn't asked the same question over and over, unless she wants to change a piece of information.

A critical limitation on relationship retailing is that you have to give the customer extra value to get extra information. If the customer does not see, early and often, evidence that the data she gives generates valuable gifts in return, then the data will dry up.

A big advantage of relationship retailing is that you can direct attention to new items. Most new products fail because they don't enter the *consideration set* of the shopper. They are never even noticed, despite the marketing done on their behalf. In supermarkets, over a month-long period, only a small percent of items purchased are different from the previous month's purchases. A smart relationship retailing system can know which new products might have the most appeal for a customer, and make a few, targeted suggestions. That way, the new product gets noticed, but the customer doesn't get annoyed.

The basic relationship strategies of cross-selling and upselling drive the process, but now they can be transplanted from slow-moving, occasional-purchase businesses like financial services, to the hotter, faster, more lucrative arena of retail.

Relationship retailing is real-time retailing

This is implied by the definition above, but it gains from being considered separately. The real-time retailer knows where things are at every moment of the day. He knows the location of each piece of merchandise: what's on the shelf, what's in the back room, what's coming in at noon, what can't be here until tomorrow, what must be restocked today and at what time. When the second-to-last item in a SKU is taken from the shelf, an associate can be alerted. Electronic shelf tags can suggest a substitute product at an empty spot until the moment the item is restocked.

The real-time retailer knows where employees are. Who is restocking? Who can be freed-up to answer customer questions?

The real-time retailer knows where customers are. He can note the entrance of a customer, summon a list of the items she always buys, and verify that those items are in stock, with an emergency alert if a critical item is missing. It's important to capture that sale, but it is far more important not to disappoint a valued customer. An absent product creates a *winback* situation. You have let someone down and may need a special effort to keep the relationship from slipping. It's much better to avoid that failure in the first place, even if somebody has to scamper.

Maybe everybody is skipping one aisle in the store. If you notice that, you can do something about it before it affects the month's numbers. Some items that appear to drive the purchase of another item are placed far from that item in the store. What can be done about that? Should there be a reminder at the shopping cart when customers appear to have bought the frosting but not the cake?

A supermarket could use the number and location of carts in motion to know how many people are needed at checkout, so they don't wait until customers are standing in line to start summoning extra help.

Relationship retail includes sense-and-respond

The meaning of sense-and-respond is implied by the definition. But if you are thinking about how to create a better experience, it makes sense to think about each part separately. Sense-and-respond retailing is the idea that integrated technologies respond to customer events automatically, as directed by business rules.

To understand the power of this, think about the retail experience today. Retail employees are managed from a distance. Managers and associates who may not be highly trained or have years of experience, make unpredictable merchandising decisions, based on obsolete information about what's on the shelf, to meet the needs of an unknown mass customer. They put a huge proportion of the inventory on reduced-price promotions that are aimed at no one in particular and are based on months-old data.[1] Sometimes they promote things the store is just about out of. This cannot help but produce an unpredictable customer experience, tolerated only because the competition is just as unpredictable.

Here's another problem. Customers interact with the retailer across every inch of the selling floor. Customer decisions are made in the middle of the store. But all of the retailer's technology and processes—all the information and all the ways to respond to it—are either at the front-end or in the back room.[2]

With RFID, you can do better. Using tags and sensors and software, you can collect data on business events, as they happen, and make decisions that respond intelligently to those events automatically and instantly, based on pre-defined rules that embody the best thinking of your most experienced merchants. You can modify those rules when customer behavior suggests a change. You can apply those rules in response to the behavior of customers whose values, preferences, and behavior you know. You can develop sense-and-respond strategies to increase short-term profitability and long-term customer satisfaction.

Think of *Total Retail Revenue* as the sum of *Traffic* times *Conversion* times *Average Sale Size:*[3]

Total Retail Revenue = Traffic x Conversion x Average Sale Size

In the current backward-looking, mass-retailing model, most marketing is focused outside the store, on primitive traffic building. In a sense-and-respond environment, you can transfer resources to inside-the-store strategies that improve the conversion rate or increase sale size. Then every customer who comes in the door is worth multiples more.

SENSE-AND-RESPOND DEVICES

Sense-and-Respond is a technocentric idea. Here are some devices that could be tied together in a retail sense-and-respond network.

Kiosks

The kiosk has been the workhorse of early smart retailing. Kiosks let VIP customers identify themselves. They let customers look up specific products with Internet-like efficiency, already a differentiating feature in bookstores. Kiosks can provide detailed product information, with competitive comparisons in categories where that is useful. Kiosks can provide environmental impact information about products. They can offer individualized coupons. They can alert a customer to sales and promotions that are aimed at that specific person's long-term value, preferences, and behaviors.

With kiosks, a customer can order customized products, from monogrammed shirts to custom-cut steaks to deli sandwiches, and have them meet him at a particular place within the store. He can order products for home delivery if he doesn't want to carry them. At Best Buy, he can ask for customer service to meet him in the home theater section, in a few minutes or tomorrow, with recommendations for a system based on a budget and a room size.

Kiosks permit customers to sample digital products, like CDs or games or movies. They can dispense digital products, which are customized to order.

Kiosks can make a service appointment, in-store or at home. If a customer brings the product with him, he can access its internal RFID tags to diagnose the service problem instantly and automatically.

Finally, a kiosk can serve as an automated checkout and payment point. Its reader can pick up all the tags on items in the customer's shopping cart and add up the bill. He taps his card on the screen on the way out the door and picks up his receipt if he wants to, or sees the information on a screen. He's rid of the checkout line forever.

Mobile shopping assistants

The mobile shopping assistant has already been mentioned. Current designs mostly feature a small black box fastened to a cart or to a basket. The mobile shopping assistant can do many of the things a kiosk can do, plus some others. It can access a customer profile and update it in real-time. It can search a store for specific products or brands. Reading tags as a customer puts products into the cart (or takes them out), it can keep a running total and set him up for instant automated checkout.

The mobile shopping assistant can look up product information anywhere in the store, at the moment when a customer wants to know. It can offer competitive comparisons, and point out occasions when a competitive brand is on sale. It can provide environmental impact information about a product at the point where he is deciding. It can summon a sales associate.

The mobile shopping assistant can deliver real-time, individualized promotions. It can offer the perfect coupon based on customer value, preferences, or behavior. It can offer a coupon as the customer approaches a product's shelf space. It can offer a coupon based on a complementary product she has picked up. It can report loyalty points, and show in real time the loyalty point consequences of a purchase she is considering.

The mobile shopping assistant can read the tag on a product she is considering, and show her a list or pictures of similar products of the same size, or in the same color or at the same approximate price point. It can show products that might go with a product she has picked up. It can offer a reduced price for combining the product she has picked up with a complementary product, the way Amazon does. It can read the tag on a food product and offer recipes or serving suggestions. It can read shelf tags on an empty space, and guide her to substitute products, or print out a rain check. It can read sensor tags on meat and produce and make a last-chance determination of freshness that is more accurate than a generalized serve-by date. Toward the end of a shopping trip, it can show items she usually buys, but hasn't picked up yet.

The monitoring gate

Basic to RFID systems is the monitoring gate, a reader that notes every tag that enters a particular space. A monitoring gate can make certain that VIPs are recognized as they enter a bank, store, or service establishment. At the door between the back room and selling floor, it can note when end-aisle displays for a particular promotion are actually deployed. It can prevent shrink, whether due to customers, employees, or suppliers. If ten tubes of lipstick start to leave a drugstore at the same moment, it can alert a nearby manager.

Because the monitoring gate keeps high-value items from walking out the door, it permits a different kind of marketing of high-value items. Right now, the curse of the health-and-beauty-aids business, for example, is that drug stores keep many of the expensive products locked up. This means customers can't handle them and decide what they want. A strong gate system lets the store be much more open to customer handling. Where stores today hang up a piece of paper and invite the customer to bring it to the register to buy a small, high-value item, future stores can put the product itself out where she can examine it, read the package, compare it with other products, and decide whether it is the one she wants.

Smart Shelf

The **smart shelf** has readers that notice when a tagged product is put on the shelf, and when it is taken off or moved to another shelf. Smart shelf is a mature technology; lots of applications have been tested. As more products are tagged, there will be many opportunities for smart shelf applications. Smart shelf was developed first for inventory control. It provides the best possible measure of how many instances of a product are actually available to the customer to buy. Wal-Mart speaks of those items which sell 6 to 15 times a day as critical to out-of-stock control.[4] Absent a smart shelf, it's hard to keep such a product in stock without granting it crazy amounts of shelf space. If a product is absent from the shelf, a smart shelf can direct the customer to a substitute product.

However, once the technology exists, there are many uses. Cosmetics have many color combinations, and marketers want to array them in a particular order. Smart shelf applications can set the threshold of disorder that summons a re-stocking associate.

Smart shelf helps stop shrink. Working with a shopping assistant, it can know if products leave the shelf, but do not go into a cart. You can set a tag to alert the smart shelf, and raise instant alarm if a product is tampered with, if the package is torn or the cap seal is broken. Some products are particularly vulnerable to theft. Certain drugs in a pharmacy, for example, are expensive and small. Smart shelf applications can know which employee (wearing an RFID badge) removed a particular bulk container and when it happened and when or whether it was returned to the shelf. This sort of prevention is vastly superior to any after-the-fact response.

Finally, smart shelf can address mistakes in merchandising. Products do not occupy too much space or the wrong space where smart shelf rules are in place.

Part of the smart shelf idea is the **electronic shelf label**: a sign showing product-and-price that can be changed instantly, remotely, by radio. Putting the right prices on the right shelf for every product in a supermarket is a grinding, repetitive, time-consuming task. A supermarket or a mass-market retailer might have 40,000 to 50,000 different SKUs. It might need to re-price 10,000 of them each week. With electronic shelf labels, you can simply update prices on a computer spreadsheet, and post them electronically. Instead of a job that uses several employees almost all the time, re-pricing can be done by one person. When prices can be changed so painlessly, new tactics become available. You could watch the prices of volatile commodities, and re-price them as the cost of replacing them changes, literally hour by hour. You could drop the cost of heart-shaped candy boxes hour by hour as Valentine's Day wanes away, to maximize revenue and minimize the amount that must be blown out below cost. You could experiment with price levels by category, and learn what the optimum price really is. It might be different at different seasons, and probably is different at different stores. Everyone assumes that the mass-market retailers have figured out how to optimize price, except the mass-market retailers themselves.

Smart POS

Smart point-of-sale displays offer a controversial and fascinating new marketing medium. Displays at the point-of-sale are powerful and underused—squeezed out by the retailer because they take up priceless space. Flat screen technology, fed by wireless networks, offers a point-of-sale message so dramatic that it can justify the space it takes up, not just to the brand but also to the merchant.

The real giant step requires RFID. With tagged loyalty cards for VIP customers, you can deliver a personalized message to an individual customer, right at the point of sale. Conceptually, this is just about the ultimate. Imagine an event in which an electronic message, with video and text and even audio if desired, speaks to the individual needs of a particular person at a particular moment in time, at a particular place in the store—*at the true moment of decision*. Show one product as this customer goes by and another product as that customer goes by. Choose from products the customer has purchased in the past but hasn't picked up this trip. Or take your cue from products already in the customer's cart on this shopping trip. Focus on nutrition for one customer, eco-advantages for another, and price-per-serving for a third. Segmented messages responding to the customer's loyalty card are programmable, and very involving.

Smart POS displays can switch from a poster that catches the customer's eye to a sheet that gives detailed information, at the touch of her finger. They can show nutritional information, competitive comparisons, or environmental impact information.

Employee work pad

The **employee work pad** is a tablet computer that translates information on goods received, products running low on the shelf, and customer-service requirements into an individualized, prioritized, instantly updated things-to-do list. A customer request, placed on a mobile shopping assistant, goes to the top of the list, and locates the customer within the store for the sales associate. Customer profiles can give the sales associate information about the customer being served, from shoe size to VIP status. An urgent restocking requirement is assigned to a particular person at a known moment in time, so each task has an owner.

The special problems of apparel retailers

Nightmare! Imagine a business in which every item you sell comes in four styles and two colors and nine waist sizes and eight leg sizes, and the customer for one of this set of nearly identical items is not a prospect for any other item in the set. Imagine that the items in this look-alike set cannot be quickly distinguished from each other, that potential customers come in every day and move the items around in unpredictable ways, and that every day you must figure out which of the various size combinations is no longer on the selling floor, and replace it from a stash of items in the back room which are equally hard to tell apart. Imagine that each of these items has a full-price lifespan of only thirty days, after which it must be marked down, week by week, to a much lower price. If you lose it in the back room for a while, you may have lost your opportunity to sell it at a profit. Now imagine that, of all the people who are lured into your store by its expensive location, by expert merchandising and marketing, and by reduced prices, *60 to 70 percent go away without finding what they were looking for.*[5] Welcome to Levi's. Or Kohl's. Or Macy's. Or most other apparel retailers.

The inventory and timing issues that face apparel retailers are truly daunting.

Consider the stuff in the back room. The description above understates the problem. If your customer summons a sales associate and asks for an item that's not on the shelf, it may be so hard to find that the associate comes back out and pretends it is out of stock. It is not at all uncommon for stores to order more of items that are already in the back room, unfound, but in ample supply.

Customers pick stuff up in one place and put it down in another, and it's *gone*, un-sellable until somebody else finds it and puts it back. (There's an RFID application with smart racks that trigger a pager alert or a phone alert when something gets put in the wrong place, so an associate can track it down and put it right.)[6]

Everything has a bar code on it. But taking inventory by bringing a reader to every shirt and slip and sock is a long hard job. It's expensive. Employees put it off. Employees cheat.

If you are a specialty apparel retailer, the problem of unsold merchandise is yours alone. You cannot kick stuff back to the brand manufacturer and say this didn't sell, as some mass-market retailers have been known to do. All you can do is keep marking it down until it goes away.

Even shoplifting is harder on apparel retailers. A product that walks away creates both an immediate loss and an undetected hole in the inventory on the floor. It may keep the store from selling that item several times before it is detected.

In this situation, a little bit of RFID makes a big difference. In some pilots, RFID tagging reduces out-of-stocks on an apparel retail sales floor by *30 percent or more*.[7] At the point-of-sale, the tag generates an invoice and updates inventory information, so the size-and-style hole in inventory on the floor can be filled quickly. Inventory in the back room is quickly detectable with a portable reader, so it's easier to replace sold items, easier to respond to a customer query about an item not on display. It's so easy to do a sales floor inventory, it's done twice a day instead of once a week.[8]

In an RFID environment, specialty apparel retailers actually have an advantage. They control their supply chain pretty much from end to end; they probably have the leverage to get tagging done at manufacturers, and don't have to deal with pushback from branded manufacturers.

Every apparel retailer's business model assumes that you can deploy new items on the selling floor quickly, and keep them deployed in every size and style until they sell out. It is important to understand that this happens *only* in stores empowered by RFID. In GAP's first RFID trial, just knowing what was on the floor produced an instant 2 percent sales lift.[9] Remember that the majority of customers leave empty handed. How many more could be converted with perfect real-time control of the product on the floor?

RFID on display

IBM's Vue/Pricer is a sort of smart shelf for apparel. It has wireless price displays that can be changed remotely. It can show a list of the sizes that are in a clothing stack, updated in real time. It alerts a sales associate when a style or size has run out. It can tell the associate or customer if the missing item is in the back room. Wow.

Paxar's Magicmirror, installed in a fitting room or on the selling floor, has the ability to respond to merchandise tags based on pre-set business rules. The Magicmirror can read an item that the customer wants, and see instantly if it's available in the different size that the customer wants, and if so, where it is. It can look at an item and offer complementary items. "Here's a top that goes with that skirt. Here's a sweater that goes with that top. Want to see some matching shoes?"

A process for selling complementary items fills a quick-search function with women who may decide that a jacket should become an outfit. Marti Barletta, author of *Marketing to Women*, says recommendations are useful and motivating to women shopping.[10]

The smart table can read the tag on an item and show that item worn by a model. Dresses, in particular, are easier to visualize this way.[11] In another variant of this model, when a customer hangs a suit on a hook, a screen beside it shows fabric information, sizes and colors in inventory, fashion show images of the product, and things that go with it.[12] There's a smart table application just for markdowns, because those get moved around most. It tracks them so associates can keep putting them back.[13]

Throughout this book we have talked about cross-sell and up-sell strategies. The reader table at retail is a simple way to automate these strategies. By making it a self-service process, you can keep the customer from feeling too aggressively sold to, and create a welcome set of suggestions, much as Amazon does online. The power of this reader table is just now being discovered. There are applications in hardware, where customers who want to solve a problem need to identify all the parts and tools required. In appliance stores, you could take a tag to the table to get features comparisons. At a service counter, you could bring in a product that isn't working for an instant diagnosis.

Changing out-of-stocks changes the customer experience

Since the beginning of retail, every serious merchant has obsessed about out-of-stocks. The problem only gets bigger in our time-starved, over-stored, convenience-driven, fuel-sensitive marketplace. Out-of-stocks annoy the customer. About half the time, they lead to no sale, rather than a substitute sale. That punishes the retailer. About half the time when the customer buys a substitute, it leads to a continuing brand change.[14] That punishes the brand marketer, who is also trying for a relationship. Yet our computerized retail environment has not succeeded in controlling the problem. Wal-Mart's suppliers have gossiped for years about inaccuracies in the inventory of the world's most data-savvy retailer, and how powerful the effects of those errors can be.

Wal-Mart has done rigorous research into the power of RFID to fix the problem of out-of-stocks. It conducted a famous matched-store study in which some stores had tagged merchandise and some did not, to see precisely how much difference RFID made.[15] More than 4,500 products were tagged. In the tagged stores, items out-of-stock were reduced by 26 percent. Products that went out of stock were replaced 63 percent faster. The total amount of in-store inventory declined despite the improved coverage. Associates who do re-stocking found they had more time to spend helping customers.

Bear Stearns studied the results of two retail early-adopters of RFID for selling floor inventory control. They saw a 9-to-14 percent reduction in out-of-stocks.[16]

One of Best Buy's RFID initiatives focused on media products—movies on DVD, music CDs, individual video games, and the like. This is a tough category for out-of-stocks. There's a huge variety of items in inventory, and only a few of each. There are lots of direct-to-store deliveries of small numbers of individual products from many different suppliers. Hard numbers have not been released, but Best Buy says its out-of-stock problems in this difficult category were solved by RFID tags.[17]

RFID and brand management

Brand marketers are one step removed from the selling floor. They desperately need to understand the customer experience at retail. They must view the retail moment-of-truth on which their livelihood depends, in an awkward over-the-shoulder relationship with the retailer. So the quantitative, real-time, automatic data that can come from an in-store RFID application is a potential godsend. Consider the things a brand marketer could now know that were simply invisible before.

Smart-shelf testing can measure which POP (point-of-purchase) designs are most effective within a store.[18] This sounds like a small thing, but is, in fact, a giant step. The deployment schedules of POP and the bar code data that reports sales contain huge time lags. In the past, it has been hard to know when a particular display was deployed in a particular store. The result is that the marketer might not know whether one particular POP execution is more effective than another until five or six months later—much too late to do anything about it. Smart shelves could answer the question in days.

With a smart shelf, it's a no-brainer to measure "put-backs," situations where a product has been taken from the shelf, examined, and returned to the shelf.[19] What a thing to know if you are that product's marketer! Maybe there is a problem with the package. Maybe there is a product fact the customer wants to know and can't find out. Maybe there is a detail that is wrong in a product that is right. A new product launch is a huge, all-or-nothing effort. If the product doesn't sell through, all you know is that something was wrong with it, and generally, the whole idea is dead. Put-backs might lead to research that allows you to fix a problem in test-market which would otherwise have doomed the entire launch.

Promotion Execution: An early success

Price promotion by brand marketers is a universal and fundamental tactic. Consumer spending on these promotions is enormous: 12-to-15 percent of annual sales for a typical fast-moving consumer goods brand.[20] In a typical execution, a marketer offers a price reduction across many different stores. The special price is advertised, perhaps with a free-standing insert in a newspaper, or perhaps in the item-price advertising of the retailer.

The brand marketer prepares special packaging or a special display for the price-reduction. This is critical to the success of the promotion; it adds extra visibility and an in-store reminder of the special price.

However, without the display, the price reduction is not nearly as productive. Without the display, the marketer may be just cutting the price and selling the same amount of product to the same customers at a reduced profit. Without the display, there won't be enough product in the store to respond to demand created by the promotion.

It matters to the brand marketer that the display goes onto the selling floor at the beginning of the promotion period. In a two-week promotion, each day is 7 percent of the total selling time. Every display that goes up late costs a lot of money. If fourteen stores start one day late, that's as bad as losing one whole store from the promotion. If the average store is even one day late, that's like losing many stores. Advertising focused on the promotion is wasted. Demand that was created by the advertising is shifted to a competitor, or frustrated altogether.

The retail store with tens of thousands of SKUs has hundreds of promotions going on all the time. They are not nearly so important to the retail sales associate as they are to the brand marketer. It's a pain in the neck finding and deploying and making room for all the promotional displays. From an employee's standpoint, no one is going to know if it was done on time. One study puts the on-time deployment of promotional displays at 56 percent.[21] That same study says 1 percent of brand marketers believe they get excellent value from the promotional dollars they spend through retailers. A more optimistic study estimated that if displays went up on the day they were supposed to, the average promotion would increase its sales by 20 percent.[22]

If you put an RFID tag on promotional displays, with a reader on the store's back door, a reader at the store's box crusher, and a reader at the door between the back room and the selling floor, you can know what's happening in real time. There is no need for a sophisticated system. Tags can be put on the promotional displays by hand.

One of the interesting things brand marketers see when they tag a promotional display is displays that move back and forth from the back room to the selling floor several times. This probably means that the display box is being used to replenish existing product on the shelf rather than set up on its own.

What you could do if you could see how promotional displays are deployed

First, you could see if you made it through that last 100 feet—if your promotional merchandise is on the floor on day one of the promotion. You could know in advance if your promotional product did not make it to the store at all.

Second, you could call individual store managers, in identified exception stores, in time to get the promotion deployed on day one.

Third, you could respond when the store manager says your promotional goods have not been received at the store yet. (The tag on the box passed the reader going into the store's back room.) You could respond when the store manager says that the promotional display has already gone up. (It didn't pass the reader between the back room and the selling floor.) With real-time data, this doesn't have to be a contentious conversation. When everyone gets used to real-time data, most of the conversations will never happen

Fourth, you could get trustworthy information about how well your promotions work. In the past, you couldn't know if problems were due to the promotion idea or to bad deployment. If you solve deployment, then you can learn from each promotional success or failure. Brand marketers believe this information improves relationships with the retail buyer. Certainly it improves on the current situation where, if a promotion fails, the retail buyer sees it as the brand's weakness, and the brand sees it as the retailer's deployment failure. Better information about how well a brand's promotions work leads to more accurate forecasting, which reduces product returned to the brand marketer.[23] That's good for everybody.

Fifth, you can see when store-to-store transfers or other diversions happen. This was invisible until now.

If promotional goods themselves are tagged, you could also see how a promotion plays out in real time—not just in weekly averages. (Did it all move in the first few days? Maybe your price cut was too big.)

Absent RFID, it is hard to see how to solve this problem. There are many services used by marketers to make personal visits to stores, but it's impossible to visit all of them in the brief space at the beginning of a promotion, when the battle is won or lost.

Absent RFID tags, you don't find out about a store that didn't receive promotional goods on time. You cannot use the point-of-sale reports of sales activity to detect an un-deployed promotion. By the time you get zero sales for the promotional product, it's too late.

PROCTER & GAMBLE
Gillette Introduces Fusion

A major advertising campaign launched the new Fusion razor on SuperBowl Sunday. Plans called for big promotional displays at 400 retail locations, deployed the day the advertising broke. To maximize deployment, P&G used RFID tags on displays for the first time. They got an instant update when displays arrived at the retailer. They could tell on Day One if displays had reached the selling floor or not. Where it was not deployed, they called store managers, shared the RFID data and asked for immediate deployment. This was a highly visible, big budget campaign, by one of the most powerful marketers in the world. It was a completely new product, with no chance of confusion between existing product and new product. On day one, P&G found that 62 percent of the retail locations had the product on the floor. By day three, after lots of interventions, 92 percent of locations had the product on the floor. The average for previous introductions, without RFID, was 80 percent by day three.[24] An industry study says that stores which deployed on day one had 48 percent higher sell-through by the end of the promotion.[25]

Eight things you can do with RFID at retail that you couldn't do before

Promotion tracking is just the beginning. Real-time data from RFID tags will transform the retail experience. Count the ways.

1. **You can make checkout automatic and instantaneous.**

   Time wasted at checkout is the single most irritating problem for the customer.[26] Instant, automatic self-checkout is feasible now. It's held up because few stores can get all their products tagged. That won't happen till tags get cheap enough, but the tipping point is not far away.

   Since 2004, there have been small retailers in Japan where the customer passes her cart by a scanner and gets a receipt without loading and unloading.[27] NCR has designed and tested an RFID self-checkout system called FastLane. It tallies purchases and disables tags simultaneously.[28] Metro, in its store of the future, has used automatic checkout systems which tally the items in the cart and collect payment via credit or debit card.[29] There are also several systems where the cart does the tallying and debiting. All that remains is the giant step of item-level tagging.

   A convenience store is testing an interim step where one thousand of the most popular products are tagged.[30] All the tagged items get rung up with one keystroke. If you bought only tagged items (which can happen at a convenience store), your cart gets rung up in one keystroke. It's an interim step, but it's already a reason to prefer.

2. **You can reduce or eliminate customer waiting time in all kinds of retail.**

   Mitsokoshi is piloting a shoe store with tagged shoeboxes. Because they can be found faster, the customer wait is cut in half. In women's shoes, this dramatically affects the number of styles that are shown to each customer. Showing more styles produces a reliable 10 percent increase in monthly sales. The system also notifies the manufacturer in real time as each pair is sold, which should improve replenishment.[31]

One of the more complicated business models in retail belongs to Pendragon, a company that rents costumes to people who attend Renaissance fairs.[32] The stores are in temporary locations, sometimes in tents. The inventory is wildly non-standardized. The costumes are multi-product combinations, some parts of which have to be kept together. People who come in for a costume are not willing to wait very long. And everything that has to be identified and itemized as it goes out the door, also has to be identified and itemized as it comes back in. The process is automated with inexpensive, passive tags. An antenna in a countertop pad reads each item going and coming.

3. You can identify and prevent quality problems.

When tags carry sensors, retailers can use sensor data to detect quality problems that have not been detectable before. They can find problems inside packages that occurred before the product got to the store, and problems that would normally require an expert.

Products that spend a very long time in storage, like good wines, run a risk of temperature variation that can ruin an expensive product. There is no way to know if it has been ruined without opening it. Tags can accompany bottles for years and provide an accurate record of their exposure to extreme temperatures.[33]

Ballentine Produce sells tree fruit and grapes. Critical to its success is getting the product out of the back room and onto the selling floor during the brief period of perfect ripeness. But Ballentine has no control at all of the process in the individual store. Ballentine has begun a program using tags like these on promotional displays, to warn retailers that it's time to bring out the product. Early indications are that its warnings measurably improve the freshness of the product on the selling floor.[34]

Shock sensors can tell when a fragile product has been dropped more than six feet, and is probably broken, without opening its box. An appliance retailer, for example, could segregate that product and check it before selling it to a customer. The tag could be read at the back door, where the retailer would have the option of not accepting that part of a shipment.

Kraft has begun a systematic program using tag-borne sensors to protect the freshness of its food products. In some categories, this may be a marketable advantage with the retailer and the customer as well.[35]

4. You can make it easier to search for things.

Today, you can go into a big-box bookstore and do a quick, kiosk-based search to find out if a particular book is in the store. If you could do the same thing at one of the big mass-market discounters, or at the giant wander-for-days home centers, that would be a competition killer. It would almost certainly increase revenues per visitor. The more time-constrained people get, the more appealing it becomes.

5. You can deliver better information at the point of sale.

Why shouldn't products carry the information a customer needs to decide whether or not she wants to buy them?

A Japanese cafeteria chain puts tags with calorie and nutrition information on individual dishes. A reader on the buffet table adds them up. A customer can assemble a meal with a real-time look at its health implications, and swap out that sundae for the fruit plate.[36] A tag on a bottle of wine could hold a weblink, which the Internet-enabled cell phone can use to go to a third-party website to read the wine's reviews.[37]

An important new way of delivering information at the point of sale will emerge as cell phones become tag readers. Nokia and Phillips have phones that can be used as tag readers. This application is popular in Japan and Korea, but has not yet emerged in the United States. Passing a phone near a tag picks up its content. Customers could pass a cell phone reader over the tag on a CD and listen to a sample.[38]

A phone tested last year in Seoul includes both an RFID-based contactless credit card and an RFID tag reader. An ABI research report says that by 2010, half of all phones will have this capability.[39]

One of the applications created by the cell phone reader is the smart poster. Smart posters show the headline and visual, on paper, simplified for dramatic effect. They can attach long copy with specifications or prices or warranties or instructions in a tag, which the viewer can read by phone. It's an inexpensive way to add information, and the information can be changed, if unlocked by its original poster. It can even be changed remotely. The NFC Forum, an industry group, has released a standard protocol for Smart Posters. Hewlett-Packard has a tag specifically designed for attaching text, audio, or video to posters. It has a big memory, a short read range, and a very high data-transfer rate.[40]

A smart poster tag can hold a preview of a movie, CD, or video game at the point of sale. It can be sampled as video and audio on a cell phone. Smart poster tags can have schedules for plays or movies, ringtones from recording artists, and computer wallpaper. The tag on a poster can hold a web link, which the Internet-enabled cell phone can use to go to a website.[41]

With a cell phone reader, information that is *inside* a package can be made accessible to the shopper; the warranties, the operating instructions and so on, can be on phone-readable tags.

When a customer brings a product back for service, a tag could carry its service history, and information that will diagnose the problem. Combine this strategy with cell phone readers, and you have the ability to diagnose a problem product from home, so you know whether it makes sense to take it back or not.[42]

6. You can leverage information as a product feature.

Since the beginning of retail, information has been a product feature. But when the label can carry the info-load of a printed book, or audio or video information, or all three, the information becomes a much bigger part of what the customer is buying. Marketers have written for at least a decade about future food packages that give directions to the microwave and clothing tags that talk to the washing machine. That might not fill an urgent unmet need.

However, there are other applications that will. When a patient picks up a bottle of prescription pills, she gets a bundle of printed data that runs to thousands of words. Sometimes drug stores make the customer sign a paper that says she received that information. Nobody who isn't a drug company lawyer thinks any meaningful communication is happening there. Yet the information, or some of it, is critical to the product's benefit. Put it on a tag that stays on the bottle, a tag the customer can read with her cell phone, a tag that has pictures, or even animation to explain what she needs to know.

Some products should be self-aware. Underground pipelines should tell someone if they are leaking, just like tea kettles announce when the water is boiling. Photographic film should tell the photographer if it's not any good any more. Tags with sensors can do that.

There's a Hasbro *Star Wars* toy whose RFID tag contains a few snippets of digital dialog. When you put it next to another toy in the set, it speaks to the other toy. It says different things to Princess Leia than it does to Jabba the Hutt.[43] (Anyone would.) *Context-sensitive information*, collected by a sensor and carried on a product by a tag will guide next steps in responsive retailing. There are point-of-sale screens (in the lab) that watch you watch them. If you squint, the type gets bigger.[44]

7. **You can personalize the in-store experience for key customers.**

In the Internet e-commerce environment, you can walk alongside the customer as she wanders through the website. You can look over her shoulder as she considers things. You can see what she looked at just before she decided to buy or not to buy. You can look at what she searched for, and how she phrased the search. You can look at how much time she spent considering a purchase that didn't get made. (Time spent is a much-studied indicator in retail.) You can look at everything she has bought before, and how long ago it was. You can look at size of purchase and whether she pops for the high-priced next-day shipping. You can see things she has put aside for later but didn't buy when she found them.

In a word, you can personalize the experience to address the long-term value and values and past behaviors of a particular individual customer.

All this is accepted in the online environment. Nobody objects. If you tried that at the Victoria's Secret store, you'd get arrested. If you told your customers at Macy's precisely what they bought last year, some of them would leave and not return.

But the wind is changing. Borders gets permission to track your purchases and rewards you for your data. So does CVS. Both of them sell products that are pretty private. So maybe it's time to start thinking about how to personalize the experience at brick-and-mortar retail for your most valuable customers.

8. You can solve shrink problems.

Loss and theft are $30 billion in U.S. retail.[45] The rise of eBay and other automatic online auctions has brought about the massive institution of e-fencing, in which shoplifters eliminate the middleman. This raises the shoplifter's revenue from about 30 percent of retail to nearly 70 percent. In 2003, the average shoplift was $263. In 2005, it was up to $855.[46] Right now, the major defense against shrink is large primitive tags on a few high-value items, and the locking-up of small high-value items. Each of these has weaknesses in defending against employee shrink, and both create a customer inconvenience. When most or all products are tagged, and an automated ID system for a whole store is in place, a chain can start the process of eliminating shrink.

Smart Vending: A new business opportunity

It's easy to add wireless communication and data processing power to a vending machine. Such a machine could respond intelligently to individual customers and to its changing situation. It could know what it's running out of and arrange to be re-stocked efficiently, the way a smart shelf can.

It could figure out which combination of products generates the fastest sales or the largest profits in a particular location. It could, of course, take credit cards. It could offer different products in the morning than it does in the afternoon. It could offer products in different sizes (juices, drinks) and create the package mechanically, immediately after its button is pushed.

The next step is a store-sized vending machine, an entire convenience store without employees. Tag tracking of products sold empowers a real-time replenishment trip every day or so, with the replenisher carrying only the products that are running low. Convenience-store-sized vending machines will be opened before this book is.

Services will be sold automatically as well. Imagine an unattended car wash. The door opens in response to an RFID tag on the car's windshield. Soap and water are dispensed automatically. Brushes and buffers whirl as the car reaches the right place. And the tag on the windshield picks up the check. Frequent users get big discounts. (Almost all the carwash business is frequent users.) This is not a vision of the future. It's working now. And it's sold as a *relationship*. Transaction customers can't even get in.[47]

Some limitations

Why hasn't more of this stuff already happened? Three factors limit the revolution in relationship retail. All three are fading.

First is tag cost. Tagging a product at retail today costs around ten cents. It's falling, but it won't disappear. If you put a five-cent tag on every product sold at retail in the U.S., the cost would be greater than the total revenue of Wal-Mart. The paradigm-changer should be printed tags, a whole new technology that will eliminate much of the cost. Massive investments are being made to develop printed tags. Their immediate arrival has been predicted for years. Here's another voice saying it won't be that much longer.

Second is the data avalanche. No business in existence today has worked with the amount of data generated by hundreds of thousands of tags, read by long rows of smart shelves, and read every few seconds, day and night. It's not theoretically difficult, but it is not a structure standing around waiting to be used.

Most important is the problem of excessive intimacy at retail. In Japan and China, some shoppers now carry cell phone straps that have RFID tags for personal identification. They do this in order to get more personalized service.[48] But the U.S. is not quite there yet. When the computer in your cart makes personalized suggestions, it will take some getting used to. It is not obvious how to phase it in. There is a danger that retailers will underestimate the rewards necessary to justify a relationship. However, there have been a lot of steps toward empowering the VIP customer, and none of them yet has generated as much pushback as relationship marketers predicted. Now comes a massive assertion of relationships. It won't go unnoticed or unresisted. But it won't go away.

CHAPTER 8

RFID and the Greening of the Customer Experience

INTERLOPERS

The Burmese python grows to 24 feet long and weighs up to 200 pounds. It has long, needle-sharp teeth, and 400 ribs, with muscles strong enough to crush pretty much any small or middle-sized mammal. It is a lightning-fast, powerful, highly determined, indiscriminate eating machine. And there are 4,000 to 6,000 of them on the loose in Southern Florida.

They came in as pets, imported from Asia. They are described as cute. You can buy them at garage sales in Florida for about $20. In its hatchling year, a baby python might be six or seven inches, but by the end of its second year, a Burmese python can be seven or eight feet long. A three-year-old python can kill a child. A 20-footer can swallow an adult male. For whatever reason, a lot of pet owners lose interest after a while and let their pythons go.

Burmese pythons are very much at home in the swamps of Florida. They eat enormous quantities of rats and mice, squirrels and birds, small deer, occasionally a bobcat. Sometimes they eat an alligator; sometimes the alligator eats them.

They eat the wood stork and the mangrove fox squirrel, which are endangered species and may be completely extinguished by the python. They compete for habitat with the eastern indigo snake, and may soon extinguish it as well. They mate in the spring and lay a hundred eggs at a time

Lately, pythons have become more visible outside the Everglades. A vegetable farm near the park killed 51 of them with farm equipment, while plowing in the spring of 2007. The Miami police pick up a couple dozen each year. People in the Miami suburbs are losing cats.

Burmese pythons are not the only non-native species to invade the

United States. In Florida alone, there are African monitor lizards, South American anacondas, Vervet monkeys, giant iguanas that can grow to fifty pounds and will absolutely eat your shrubbery, and a very large number of Asian and South American tropical fish. The U.S. Fish and Wildlife says that invasive species are the single biggest threat to U.S. eco-systems. Government estimates of damage from invasive species top $100 billion a year, mostly from fish, which may, in the long term, be the most important. But pythons are the only ones that will eat you.

It has been proposed that importing pythons be outlawed. People who study the problem say that this would merely move the trade underground. Many containers of exotic animals are shipped into the U.S. every year. No one opens them. No one inspects them.

Now it's proposed that we inject imported pythons with RFID tags. The tags could work two ways. Captured females could be tagged and released in the spring, as a way of leading wildlife managers to the males, so that both can be culled and the crop of babies reduced.

You could require people who buy a python as a pet to have it tagged. Avid makes a tag that is small enough to be injected. Seven dollars for the tag; $40 for the veterinarian.[1] That tag might help to track down an improperly released python and it would identify the owner who let it go. It may be a small enough burden that it would not drive the pet owner underground.

This might not sound like the definitive solution. But, in the history of the United States, no exotic species that became as established as the python has, has ever been completely exterminated. If the Burmese python can be controlled, restricted to its current geography, its numbers reduced, its power to wipe out other species thwarted, that might be the best that can be hoped for.

To protect our planet, we need to live differently than we have in the past. We need changes in the things we do, changes in the way we do things, changes in the things we pay attention to.

There is not much agreement about what must be done, or who must do it, or even about how urgent our problems are. There is not broad understanding of the underlying science. There is legitimate debate about almost every aspect of our ecological plight. And every potentially helpful action has economic consequences, which press more severely on some people than on others.

It will be difficult to get as much done as *must* be done, through top-down regulation. We need to find out whether citizens and companies can accomplish some of the things that regulators cannot. We need to find ways to empower earth-friendly choices. This chapter will demonstrate that RFID offers unique and important ways to get this done.

The customer is already there

Poll data shows a growing focus on environmental issues, and a somewhat slower-growing willingness to do something about them. A Stanford/WaPo/ABC poll says 94 percent of Americans are willing to make lifestyle changes for ecological goals. Ninety percent say they are recyclers. Seventy percent say ecological concerns change product choices. Thirty-five percent say ecological issues have affected their choice among political candidates. Thirty-five percent express concerns about global warming. That's up from 29 percent a couple years ago, but still far lower than numbers in Western Europe, for example.[2]

Actual behavior is a better measure, and there are lots of behaviors to look at. When earth-friendly products are properly marketed, they often show a competitive advantage. The marketing part is not simple. Consumers may assume that a green product will require some kind of sacrifice. It will be more expensive, less effective, inconvenient, or lower in performance. Reassurance is necessary. However, Tide Coldwater succeeded as a believable energy-saver. Planet laundry detergent has prospered as a true biodegradable. Toyota's Prius sold well enough to spawn a generation of imitators. These are just a few examples.

What happens when green becomes convenient?

There are two ways to change the way people behave. You can *push* the behaviors you desire, with incentives or penalties. Or you can *pull* people to the behaviors you want, by making them easier, and more convenient. A lot of people drink instant coffee, but not because of how much better it tastes. The effect of tagging environmentally sensitive stuff is to make it more convenient to do something about it.

The other consistent learning from eco-poll data is that people in the U.S. are not yet prepared to make large sacrifices to meet sustainability goals. This may change, but it has been stable for a long time. Answers to that Stanford/WaPo/ABC poll emphasize that Americans want government policies that fight global warming to be low-cost policies. Policies that would produce a large increase in home electric bills, for example, are not seen as appropriate. But policies that would require smaller sacrifices generate enthusiasm. An important dimension of RFID in the drive toward sustainability is its potential for incremental change without wrenching personal sacrifice.[3]

Visibility changes everything

If you can see where stuff is, consumers who really want to, can change the way they deal with it. They can, for example, change the waste stream. Batteries are dangerous? Put a tag on them. Make it easier to track them down. Fluorescent bulbs are poisonous? Put a tag on them and keep them out of landfill. Recycling is too time-consuming? Use RFID tags to make sorting automatic.

With RFID, you can put eco-information at crucial decision points, on the retail shelf for example. You can offer new kinds of incentives for earth-friendly behavior. You can track environmental hazards and remove them. You can increase people's ability to do things remotely, and thus reduce emissions. You can create a responsibility on the part of people who manufacture things for retrieving those things when their usefulness is gone. You can do a better job of re-using things by identifying and tracking their reusable components.

Some of this is just little improvements to existing processes; make it easier to sort recycled goods and the garbage man will do a better job. Some of it is doing a better job with the most urgent parts of the problem; creating systems to keep track of poisons, batteries, mercury and the like.

However, some of it involves fundamental changes in the way the world works. There has been little connection between the act of manufacturing something or buying something and the responsibility for that thing after its usefulness is done. That is partly because of the inability to track things. They are made, they are sold, they go out into the world, and disappear.

Asking manufacturers or retailers to take responsibility for the products they send into the world when the useful life of those products is over, has been a dream of ecologists for decades. Now that you can tag and track those products, it may be an actual practical strategy.

Following are some early applications of RFID technology to environmental issues. Like the other applications chapters, the goal is to offer thought-starters. It is not claimed that this is an exhaustive list, or that all of these applications will be successful, or that all of them will turn out to have major ecological impact. As with all emerging technologies, some uses will flourish and others will perish. Perhaps in looking at what others have done, you will see opportunities for your own business.

Recycling credits and incentives

Songdo City in Korea is a nexus for the exploration of new technologies, the testing of big ideas. And one quite small idea. If a consumer drops an aluminum pop can into a public bin, her recycling account gets credit for it, automatically.[4]

Deposits for container return have a mixed record in the United States. States with container deposit laws have much higher recycling rates. But 39 states have not yet passed "bottle bills." The Container Recycling Institute says beverage container recycling across the whole U.S. has been declining steadily since 1992—down from 53 percent to 33 percent by 2006.[5]

Managing deposits is a cost-source for retailers, a small but regressive sales tax. This seems like a place for automated data collection. If tag costs fall far enough, it might be practical to collect a small deposit on many recoverable materials.

California has one of the most successful "bottle bills." Yet *one billion* empty water bottles show up each year in California residential garbage. That's one reason for the importance of a new business launched by S2C-Global and Northern Apex in 2007. Giant vending machines in the parking lots of supermarkets sell 5-gallon jugs of water, paid for with a credit card. The jugs are RFID-tagged, and when they are returned to the machine, the customer gets back a deposit, added to his credit card balance. No human labor is involved at either end of the transaction, except in filling up the machine with new jugs. That's what makes it affordable and that's what makes it an interesting model for a lot of other ecologically sensitive consumer products, like paint.[6]

Each year in the U.S. about 64 million gallons of paint are discarded, mostly in partly full one-gallon cans. When it is identified in the waste stream, it costs a city about $8 a gallon[7] to dispose of it. The consumer has no incentive to handle it carefully; much of the discarded paint is probably concealed inside bags of residential garbage. But paint in the landfill is a problem. It contains chemicals which city water treatment plants cannot remove.[8] It makes sense to tag cans of paint, to collect a deposit that is big enough to affect behavior, and to refund the deposit, automatically, when the paint cans are returned to a bin. It makes sense to ask paint dealers to administer such a process: paint has vastly more impact than beverage containers.

Saar and Thomas have proposed a process in which many hazardous products can be tagged and turned in for a deposit. The amount of the deposit would be based on the environmental impact of the product. Maybe batteries and old computer monitors would carry a big deposit, fluorescent light bulbs and empty pesticide cans would have a middle-sized deposit, and aluminum drink cans would carry a tiny one.[9] With a process like this, it might be possible to create a residential scavenger business, perhaps with a sort of kiosk that issues a credit at the curbside, when you toss in tagged recyclables.

Consider the problem of consumer electronics. It will always be easier to abandon some techtrash in an alley. Some states have laws saying that retailers must take them back. Some states do not. Staples takes back the products that it sells, but charges a $10 disposal fee. Dell will take back its used-up products if you ship them. Best Buy has bins at every store to accept no-longer-used consumer electronics. When they deliver new appliances, they take away your old appliance at no cost. The company pledges to do what it can to retrieve and recycle usable parts. Best Buy gets some participation just by making the bins available.[10] But some sort of incentive, driven by a tag that assesses the environmental impact of the product, and pays back a deposit of that amount could change customer behavior and make the recycling of consumer electronics the rule rather than the exception.

Fluorescent lights are good for the environment. They last longer and consume less power. But each one contains a significant amount of mercury, which is a neurotoxin. There are political demands to require the use of fluorescent lights instead of incandescent bulbs wherever possible, but there are not many mechanisms in place to keep them and their mercury out of the landfill. A small deposit, a tag on the bulb, and a kiosk that issues an automatic credit would help.

We throw away about 300 million tires each year. Programs that make secondary use of tires are not nearly big enough to use up this volume.[11] Tires are a huge share of illegal dumping. They create eyesores and mosquito hatcheries. Tires are already tagged in a widely-used application to alert the motorist to slow leaks, and in theft-prevention programs in Mexico. It would certainly be possible to create a tag-and-deposit program in which it makes sense for the customer to return tires to a tire retailer.

There is a recycling program in Germany that records the purchaser of targeted products, and traces who among the purchasers has recycled its container and who has not.[12] Today the U.S. consumer would probably consider this invasive, but maybe not tomorrow.

RecycleBank is a private business that pays people to recycle. Each house gets a recycling bin with an RFID tag that identifies the family. When the recycling truck comes around, as it lifts the bin, it weighs the recycled material and associates the poundage with the "account holder." Account holders get paid in coupons, donated by eco-conscious firms like Kraft, Target, and Whole Foods. Members can earn as much as $400 worth of coupons a year. The concept works even in low-income neighborhoods, where recycling is not otherwise common.

The more recycling, the less trash goes into landfill. RecycleBank gets its revenue from cities and towns whose landfill fees are reduced by recycling. The company collects a negotiated percentage of the savings.[13]

The city of Hampton, Virginia, offered incentives for residential recycling. They began with a volunteer program, such as many cities have, in which people could separate out aluminum, glass, plastic and paper, and put them in bags that are visibly different from the residual garbage bag. There was some participation in the volunteer program, but the city faced high costs and wanted to increase recycling. So they adopted a reduced garbage-collection fee for households that recycle at least twice a month, and a higher fee for non-participants. At the start of the experiment, they charged recyclers $1.52 per week and non-recyclers $4.80 per week. RFID tags were used to associate recycling with individual households. (Bar codes were tried, but they did not remain readable in the garbage environment.) By the end of the test period, the new program had created a 60 percent increase in recycled materials recovered. The city recovered its RFID tag investment within two years, from the higher fees charged to non-recyclers.[14]

RED-BAG WASTE
Lock it out of the landfill

Red-Bag waste is the tiny fraction of hospital waste that is dangerous to people or other living things: blood, blood products, body fluids, infections cultures, contaminated bedding, contaminated "sharps." *It must be incinerated.* It cannot be combined with trash that goes into the municipal solid waste system. There are well-established sorting procedures for what goes into the Red Bag. But when a Red Bag goes into the compactor by mistake, that's a problem.

Compaction crushes the Red Bag and distributes the biohazard throughout the bin. Now the area must be cordoned off, environmental regulators notified, and the whole load must be incinerated. A load that should cost $800 to dispose of may cost as much as $11,000 to dispose of. Penalties and compliance issues make it difficult to know just how often this happens. But it happens. And the cost of a single incident would nearly equal the cost of an RFID defense.

Tags on the Red Bags and antennas in the chutes that lead to the compactor can prevent the problem. One such system, designed by Sonrai,[15] will shut off the compactor to prevent contamination, and trigger both a voice message to the operator and an email to the administrator. Knowing when it happened and where the bag came from, (by reading the tag), the administrator could identify the process problem or the person who needs more training.

That same system could summon a truck when the compactor is full, rather than at a fixed interval. Sonrai says this can reduce truck visits by nearly a third. It could see when a tracked asset, like an EKG cableset gets accidentally thrown away, and rescue it. It could note, and help retrieve patient records, when they find their way into the compactor. All these would generate a savings. But the big savings happens when you lock a Red Bag out of the landfill.

Other cities have launched pay-by-the-pound systems to provide an incentive for waste reduction.[16] RFID tags on garbage cans provide a practical way to assign costs to customers. Garbage cans can be weighed by the devices that lift them and dump them into the garbage truck. The data collected goes to the software, which generates invoices.

Dial up the green message at retail

The Magicmirror, which uses RFID tags to show product benefits at the point-of-sale, is a natural for ecological comparisons. There are several independent, non-commercial eco-label and eco-certificate programs, which certify that products in a particular category meet certain standards. Others compare products and score them on sustainability issues. It's hard to fit this on a conventional package, but easy for the Magicmirror to dramatize it. Once such a vehicle exists, it becomes a critical part of the competitive struggle, and forces every marketer to be a green marketer, or at least a little less of an earth-scorcher. MediaCart could also make sustainability comparisons. Down the road, flexible, printed screens will provide a method for putting simple video messages on a product, if the message is worth the cost. That's a way to make a green message catch the eye at the moment of truth.

Relationship marketers have demonstrated that designating some customers as VIPs helps build a relationship. With proper incentives, it can be enough of a relationship to get people to carry an RFID loyalty card, which enables many retail relationship tactics. Now consider the possibility that enthusiasm for ecological issues can be the basis for a relationship. The people who care about such things care a lot, as Patagonia and Whole Foods have proven. A card that empowers alerts on eco-advantages and green market promotions might be an important and even differentiating benefit to some customers. Key to a relationship, it creates value in both directions. It helps marketers know what to say that is important to this particular customer. It helps customers know how to capture a particular kind of value.

How much can packaging be reduced?

Manuals, directions, legally required information, service information, "documentation," guarantees, and the like add inches of paper to hundreds of thousands of cell phones, printers, computers, flatscreens, and consumer appliances. The usefulness of all that paper is limited by the fact that most of it will be needed only at some indeterminate later time. By the time it is needed, almost certainly, it will be thrown away, lost, or inaccessible. The trees that were cut to produce it were wasted. All this information can be stored digitally on a tiny RFID tag permanently attached to the product, or its package, and readable with a cell phone or PDA or laptop. If you can find the appliance, you can find the information.[17]

Packaging designed to prevent theft creates big, inconvenient, and ecologically unfriendly hunks of future landfill. However, 9 percent of DVDs are stolen and retailers need to prevent theft. A new application for DVDs will put on the shelf a disc that is unplayable, effectively empty. At the point of sale, an RFID transmission enables the disc to be played. The application records the moment at which the transaction was made and the moment at which the disc was enabled, to prevent after-hours enablement by employees.[18]

Attach recycling information to products in the world

Printed labels are already overstretched, and durable products may have no place for them. But RFID tags, embedded in a tiny portion of the label, or on the product itself, can carry information that changes recycling behavior. For hazardous products, or products that must be recycled at a particular place, or products that must be recycled in a particular way, tags can provide instructions.[19]

They can also explain the urgency of separating a particular product from residual garbage. It's not at all clear that people know what happens when you dump a computer or a paint bucket or a fluorescent light into the garbage. Tags can provide detailed information, with graphics to make it urgent. Read by a cell phone or a PDA or a laptop, tags can speak to those customers who care about the issue, without taking up space on the label or detracting from the marketing performance of a package and without trying to squish the message into a compromise-sized space. Right now, you can look up such information online. But it is complicated and time-consuming. It violates all the e-commerce best practices about how many clicks you can ask people to make to find what they are looking for. Put the story on a tag.

Supply chain efficiency as a method of conservation

Wal-Mart says that out-of-stocks make people take extra shopping trips. About half of these out-of-stocks are due to inventory inaccuracies. Solving these with RFID, the company says, will not merely make customers happier, but will also save gasoline.

Products sitting around as safety inventory represent a massive amount of energy consumption that may or may not be wasted. RFID systems that slash safety inventory with responsive restocking are also energy conservation systems. David Pimentel at Cornell University calculates that growing, harvesting, processing, and transporting a pound of lettuce in California to be sold at retail in New York, consumes 4,800 calories of petroleum or other fossil fuel. The closer a retailer can come to real-time replenishment via RFID tracking, the less fuel wasted.[20]

Substituting digits for driving

A tighter management of the driving done for business purposes could both reduce emissions and cut fuel usage. Organizations like Telargo[21] are designing ways to use sensors and communication to manage driving more efficiently. Tags with sensors ride on systems within a vehicle and deliver real-time alerts to an onboard computer, which communicates with a company's data center. Doing small things right can produce large savings. Sensors alert for low tire pressure, which increases fuel use. They try to produce highway speeds very close to 60 mph, which is much more efficient than higher speeds. They provide data for evaluating the efficiency of alternate vehicles. Precise measurement of weight carried can optimize fuel use. Where engine idling is used to power heating or entertainment systems, a manager may propose battery solutions. These are not complex or expensive ideas. They are the consequence of cheap real-time measurement, and if widely adopted, can reduce car-created damage to the earth. Most important, they are fore-runners of optimization systems for consumer use.

Automatic sorting makes recycling cost less

Recycling plastics is a maddening problem. Lots of municipalities ask people to do it, but it's just eco-theater. If you mix and melt all the kinds of plastics that show up in the waste stream, you get a near-useless compound which can be ceremonially manufactured into park benches and parking-lot bumpers, but which doesn't impact the need for new raw-material plastics. Hand sorting is unreliable and grotesquely expensive. Now comes a pilot test which has successfully used cheap-and-simple RFID tags to sort plastics automatically. It's a multi-step process: a production line pulls out one valuable type and lumps all the others together. Then it pulls out another valuable type and lumps all the others together. And so on until only worthless plastics are left. In tests, it reliably generates a series of pure waste streams of specific high quality plastics, which can be used in new manufacturing just as if they were raw-material plastic polymers. It might not sound important, but this is truly garbage into goods at a cost competitive with virgin polymers. It is a good day for the planet.[22]

Optimizing the disposal process

It's not the part that you see in documentaries, but a lot of ecological sustainability has to do with the efficient handling of garbage. It's a labor-intensive, widely dispersed, hard-to-supervise business, so RFID can help.

One basic system tracks productivity of individual trucks, moment by moment. When a bin is emptied into a truck, its tag is read. Tag data is sent via Bluetooth to a sort of single-purpose cell phone in the cab. The phone uploads data to a municipal server with a timestamp. With this, central supervisors can compare the productivity of different crews. They can isolate and understand and remedy productivity problems across a fleet of trucks just as you would isolate and understand and remedy productivity problems on an assembly line.[23]

Replacing products with services

Remember that the applications described in Asset Tracking (Chapter 6) almost all have profound ecological consequences. It is a basic maxim of sustainability that when you create a service in which many people use and re-use a particular product, you have accomplished recycling many times repeated. The car-sharing business model is described in many articles on ecological science as if it had no other advantage.

Minimizing the impact of pest control

The most important rule in pesticide management is to limit in every possible way the amount of "active ingredient" put down. For decades, row crop farmers in the U.S. covered their fields with pesticides based on where they had pests in the past. Today, the rule is "scout before you spray" and spray only where you find an infestation. RFID permits a variant of this procedure in residential termite control in the southeastern United States.

Where termites are a serious problem, exterminators plant traps in a circle around a building. The traps have bait that termites can carry back to the nest. But they also show the direction the termites are coming from, so exterminators can track down the colony. It's a labor intensive business, walking a lawn and checking all the traps each week, to see which ones have termites. A new approach uses a termite-tasty "fuse" in each trap. When it's chewed through, it triggers an RFID tag. The exterminator carries a reader. He knows which traps have termites, and visits only the full ones. This doubles the number of buildings he can visit in a day, which cuts the cost of control. And pesticide only goes where the termites are.[24]

Tracking animal populations in the wild

American elk used to range all the way east to the Mississippi. Today, they have moved West, to a more sparsely populated habitat whose eastern boundary is in Colorado or western Nebraska. It is not a simple thing to foster a population which cannot be confined, cannot be examined up close, and which struggles constantly to grow its population in a habitat that is more and more closed in by humans.

One current problem with the great Colorado elk herds is Chronic Wasting Disease. It's very serious. It is a variant of the transmissible spongiform encephalopathy (TSE) that is called Mad Cow Disease. Mad Cow Disease threatens humans with variant Creutzfeldt-Jakob Disease, a horrible, painful, and invariably fatal brain disease. Chronic Wasting Disease, which affects deer and elk might be transmitted through contact with feces, urine, or saliva of other animals in the herd. It is not known whether it can spread from deer or elk to cattle, so this is potentially a life-or-death issue for the elk herds and deer herds of Colorado, Wyoming, and Nebraska. If a disease that is epidemic among deer and elk herds comes to endanger the food supply, there will be pressure to take radical and irreversible steps against deer and elk populations in the wild.

It is important that outbreaks of Chronic Wasting Disease in Colorado elk herds be contained, whatever it takes. It is important that contact between TSE-infected elk and cattle be minimized. It is important that recreational hunters of elk do not interact with potentially infected animals, and do not consume those parts of the animal which put them at risk for variant Creutzfeldt-Jakob Disease. Absent RFID, society has no easy way to accomplish this.

It comes down to this. If a group of animals is infected with Chronic Wasting Disease, that group may have to be eliminated or isolated. It will then be necessary to decide which *other* groups of animals have had so much contact with that infected group, that they must be considered either infected or endangered. RFID tags can provide a sense of which animals went where, of who their *herdmates* were. Wildlife managers in Colorado have been effectively proactive, testing the use of RFID tags to see which elk could have been infected by which other elk. The program is still a pilot. It may not be continued, but without it, there may come a point where some very cruel decisions have to be made. It is at least possible that RFID tags will preserve the life of the Colorado elk herds while the scientific community acquires the knowledge necessary to defend against TSE.[25]

The White-lipped Peccary is a large and fierce wild pig, in Central and South America. It's not yet endangered, but has come under pressure as human encroachment shrinks its habitat. It's a critical prey species that keeps the jaguar alive. One issue key to the pig's survival is a political hot potato: will the peccary endure as more and more of the Amazon rainforest is developed into farmland? Peccaries have to keep moving; their herds clean up nearby food and forage and they must move on to find more. How big a range do they need to survive? Some are now being tracked with RFID tags. Tags are fastened to tranquilized peccaries at salt licks. Readers at the salt licks and at watering holes and other places, note their comings and goings and try to learn how widely they must travel to eat. There is not another way to accomplish this. The pigs will charge an intruder, and the whole herd works together. Sometimes they kill jaguars. Scientists will learn their range only with RFID.[26]

You can inject RFID tags into fish. One early application involves tracking fish as they move upstream and pass around or through river dams.[27] There is perpetual controversy between people to whom dams are important and people to whom fish are important. It is fueled by bad information from both sides. Tracking individual fish in large numbers, possible only with RFID technology, can shift the discussion out of anecdote and into the evaluation of objective data.

Measuring habitat effects that could not be measured before

The absence of consensus about ecological issues is perhaps the most dangerous risk that confronts us in protecting the habitats on Earth. Despite talk of the "scientific consensus" behind this doctrine or that, there is not very much about ecological and sustainability issues that everyone accepts. Some of this is a problem of belief systems, of people finding it difficult to accept ideas that challenge their values. But part of it is a problem of measurement. A habitat is a difficult thing to comprehend. How changes in this or that factor might change the behavior of people or plants or animals in a habitat has been hard to know, because it is hard to measure.

Mesh networks of RFID tags and sensors create the power to measure, record, and communicate dimensions of a habitat in ways that were unaffordable or impossible before. Tags that use solar power make long-lived mesh networks possible. Scientists have used mesh networks to identify microclimates—to track miniature weather fronts that travel up and down the trunk of a single giant redwood. Measurements of light, temperature, and wind, taken with mesh networks, and associated with seabird nests, create a clear picture of what matters to the bird and what doesn't.[28]

Product life cycle management and the idea of technical nutrition

In the book *Cradle to Cradle*, McDonough and Braungart propose as a goal "technical nutrition," the idea that a product at the end of its lifecycle should be divisible into materials that can *efficiently* be used to manufacture something new, and that materials that impede re-use should be systematically engineered out. The fibers in a nylon carpet can be melted into the precursors of new nylon fiber for the next carpet, and so on.[29]

This is mostly about planning ahead. The steel in used-up cars cannot be re-used to make new cars because of the way it was painted. Different coatings could make it efficiently re-usable.

However, it is also very much about data management. If you can know exactly what a component is and how it was built, you can know how to manage the end of its lifecycle. Attaching complex data to physical objects, so that it can be efficiently retrieved a long time later is a logical application for RFID.

The problem gets new urgency from the European Union's End-of-Life Vehicle Directive, which imposes a legal requirement on car makers to *take back* vehicles they have manufactured at the end of their useful lives, at no cost to the user. Manufacturers are required to reuse, recycle, or recover 95 percent of the vehicle's materials. This may or may not be possible. The law will cause some companies to try. Dismantlers will find tags on individual components, tied to a Product Lifecycle Management database. It describes what the components are made of and the best way of recycling them known at the time of manufacture.[30] Car makers are required to pay the cost of disposing of materials which cannot be recycled. Either they will become adroit users of component tagging to cut the cost of recycling, or they will become, as builders of private aircraft became in response to regulation, the manufacturers of extremely expensive vehicles in a very small market space.

An important decision in the lifecycle of electronic products is when to repair a product. When is it worthwhile to replace a component rather than throwing the whole product away? We are efficient manufacturers and inefficient repairers, so mostly we throw things away. According to Dr. Duncan McFarlane of the Cambridge University ID labs, this too is mostly a data problem. Components have numbers, but the number on a component may change from application to application. The numbers are hard for a repairperson to locate and they can't be located at all unless the product manufacturer and the component manufacturer cooperate. So it's hard to replace a broken component. The information eco-system around RFID provides a push toward unchanging numbers and a universally accessible database. Tags within products can identify their components, which might change even within a particular model, from version to version. Other tags, at critical points within a system, may use sensors to provide diagnostic information. Sounds complicated, but the result is pretty simple: a dramatic escalation in the number of cases where a product can be repaired. The complexity and fragility of the personal computer has brought back the once-dead business of electronic repair. RFID tags may help it to grow. [31]

THE CURIOUS PROCESS OF WILLIE CADE

Seventy million personal computers were built on this planet in 2007. Five hundred million computers in the past decade—that's up from 100 million in the decade before. Ninety percent of them are here to replace computers we threw away.

That's a lot to throw away. First because each of them contains a batch of pollutants and poisons. Toxic mercury. Bromines. Fluorine. Large amounts of lead.

Second because each is full of scarce resources, dug from the earth's depleted store and used for a very short time. Copper. Platinum. Palladium. Silver. Gold. We blew up and dug up about ten tons of earth to get each ounce of those metals. Some will be recovered, most will not. It takes a thousand gallons of water to make each one of those computers.

Third because half a billion computers make a pretty big junkyard.

Almost all of them still work just fine when they are discarded—or at least, the hardware does. Maybe the operating system has slowed down and sputtered or even died of an accumulation of digital hiccups and errors and viruses. But the machine itself works fine. The life of a computer is measured by desirability, not functionality. You get to the point where you wish you had a new one, smaller or faster or with different features built in—more up-to-date. And it's a daunting task to reformat and replace your operating system. So much cost, so much hassle, you might as well buy a new machine.

Or you change jobs. You can't take it with you and the person who gets hired to fill your spot will want a new one.

Someone could recycle the materials, but they won't get much. Discarded computers and monitors by the pallet-load get wrapped in plastic and shipped in containers to cheap-labor countries to be "parted out" and recycled. That's a lot of fuel for very little return.

No, the answer is not recycling, but re-use. You want a new computer, and somebody else wants your old one.

Willie Cade sells computers to schools. They are slick, shiny, name-brand computers—laptops mostly. They are good as new and look like new. At $150 a pop, if you're a school or an at-risk child and with a three-year warranty, yet, it's a good deal. How does he do it?

The computers are donated, by companies or by individuals and trucked to his workshop in industrial Chicago. The OS comes from Microsoft—also donated, as long as the computer goes to a school or an at-risk child. He adds an open-source office software product, reformats the hard drive, cleans them up, fixes them up, sometimes juggling parts from three computers to make one good one.[32]

Willie Cade and others like him are the smart way to handle the torrent of discarded computers.

Thirty-some of the fifty states have enacted the dumb way.

"Producer Responsibility" is one of those efficient political terms that can wrap up a big lie in just a couple of words. The idea in Producer Responsibility is that manufacturers "are responsible for" that portion of the e-waste that has their name on it. The problems with this are pretty easy to imagine. First, "Producer Responsibility" just means "producer pays." They have to cover the disposal cost, but disposal is cheap. They do not have to take responsibility for building longer lasting computers, or more repairable computers, or less toxic computers, or computers with an OS that endures. There is little incentive to do any of these things.

Perhaps 40 percent of the e-waste is "orphan products." The producer is not around any more. New companies get a free ride for a few years. Short-lived companies may become the norm. But the real blindspot: our problem is not the cost of garbage disposal, it is how to prevent garbage.

What would keep computers from turning into garbage? Start with auto-sorting.

Pop the cover off a two-year-old laptop, and you have no idea what you're looking at. The motherboard might be any of a zillion versions, changed every time a supplier offered a transitory price advantage. Maybe you could drop that circuitboard into another laptop of the same model, but maybe not. Tags could tell you.

The plastics in all this stuff cannot be mixed. They cannot be gravity-sorted or optically-sorted, but they could be sorted into pure streams of valuable, reusable materials if the bigger hunks were tagged.

Incentives for the producer to make re-usable components would keep computers from turning into garbage.

Right now, the world is shifting to putting small amounts of computing power into all kinds of appliances and systems. It is absolutely possible to turn the parts of too-small, too-old computers into digital appliances. Sweden has demonstrated that digital feedback on your heating system, moment by moment, makes you use less energy and be happy about it. [33] The guts of such a product get thrown away every day, because they are needlessly difficult to salvage and re-use. Sub-systems that are pre-identified and tagged to identify themselves make re-manufacturing efficient.

Let's make e-waste into e-components. Attach the necessary information to the pieces of a computer, with tiny, inexpensive, automatically read RFID tags, and let re-manufacturing take flight.

CHAPTER 9

RFID Authentication and Product Safety

TRACKING MAD COW DISEASE

Most people know that there was once a brief and tiny outbreak of Mad Cow Disease in the United States. Not everyone knows that the problem remains unsolved—that the response by the United States is seen by much of the world as insufficient and dangerous, and that several countries have barred or severely regulated imports of U.S. beef because they are not willing to have their citizens so poorly protected from Mad Cow as Americans are.[1]

Mad Cow is just about the ultimate food safety nightmare. People who consume infected beef tissue may contract variant Creutzfeldt-Jakob Disease (vCJD), which is a horrible and fatal brain disease much like Mad Cow. VJCD cannot be cooked out or irradiated out or sterilized out with chemicals. You cannot test packaged meats for it. You cannot test an animal without killing the animal. Infected meat can infect other meat in the packing process.

When a cow in Mabton, Washington, was discovered to have Mad Cow, it had already gone to slaughter. A lot of beef was destroyed. A lot of cattle that may have been in the same herd were destroyed. But the truth is, the herdmates were never identified.[2]

If tomorrow, the same thing happened again, almost certainly the same failure would happen again. There are 96 million cattle in the U.S., at 800,000 different places. They are widely dispersed, but 90 percent of them pass through five locations during the production process, rubbing shoulders and swapping spit.[3] In the EU, where a much larger outbreak occurred, they test one slaughtered cow in every four for Mad Cow. In Japan, they test every single animal. In the U.S., we test one in ten thousand.[4]

We could track the U.S. cattle herd with RFID ear tags. The process is tested, efficient and relatively inexpensive. By one estimate, less than

5 percent of U.S. cattle are RFID-tagged today.[5] A national RFID identification system is mandatory in Europe, Canada, Australia and Brazil. It is voluntary in the United States. The next significant outbreak of Mad Cow in the United States will have an impact like a bomb in a city.

RFID can label and track not just a category of products, but every instance of that category—every package of lamb chops, every bottle of amoxicillin, every seven-year-old on the way to school, if the child wants to be identified. There are enough numbers in the EPC code to issue one for every grain of rice.

The ability to tag the individual instance of a group implies the ability to authenticate—to tell whether a particular member is or is not what it says it is. You can track the lamb chops from the farmer to the supermarket checkout, track the amoxicillin from the manufacturer to the prescription counter, and know which seven-year-old has missed the bus.

There are three activities where authentication can confer enormous benefits:

- One is human identification, discussed in Chapter Five.

- The second is the prevention and detection of counterfeits.

- The third is the preservation of product safety.

Product authentication

It is both easy and important to verify that a product is what it says it is. An RFID tag on a product is powerful and inexpensive authentication. It is as unique as a fingerprint. It can be cryptographically secure. It can address problems with counterfeits, with substitutes, and with generics. It is too expensive to replicate illegally.[6]

The easiest way to authenticate things is with some kind of secret data: a password, a code message with a shared key, or secret data combined with message data in some pre-determined algorithm. The sender and the receiver must also have a way to make sure that a message is not a replay of a previously sent legitimate message, and a way to make sure that a message has not been tampered with.

The information that is necessary to authenticate a product can be carried on the product on an RFID tag, but not all RFID tags can be used for authentication. You need a tag that cannot be correctly duplicated by an outsider. If the tag cannot be duplicated without alerting the shipper, and the container cannot be tampered with without alerting the shipper, then you have a system that can authenticate the product.

The problem of counterfeit drugs

Americans fill almost four billion prescriptions a year,[7] and more each year as the population ages. It's a big ticket transaction—almost always more than $50, counting both the co-pay and the insurer's payment.[8] Most of the time, what people get is exactly what they paid for. Sometimes, maybe five or 10 percent of the time, what they get is phony medicine. That's doubly dangerous. It causes the patient to skip a dose without knowing it. And the fake may have dangerous ingredients. A batch of counterfeit Plavix found in 2007, actually contained cement powder.[9] There is no infallible way to look at your pill or medicine and know whether or not it is real. Counterfeits have been a significant part of the prescription drug business in the U.S. since at least 2000. It's organized crime, just like the addictive drug business. The profits are up there with the profits from narcotics.[10] Everybody knows about the narcotics business, but patients and their doctors don't think much about fake prescription drugs.

It's a bigger problem than you might imagine. In 2007, Lipitor was a $3.4 billion dollar pill. You can't feel it working. You can't tell if the pill you took this morning was the real thing or not. Katherine Eban, a *New York Times* reporter, tracked down an incident that involved 600,000 doses of counterfeit Lipitor, one of many that she found. Another was a cancer medication, a late-stage life preserver with enough counterfeit doses to treat 25,000 patients for a month. In 2005, a single organization put $50 million worth of counterfeit Lipitor into the U.S.[11] In the first half of 2006, when Pfizer began tagging Viagra, they found 4.9 million counterfeit pills in the supply chain.[12]

A medicine in liquid form might simply be diluted or "cut," the way cocaine and heroin are cut. How would the patient know? The World Health Organization (WHO) says that 25 percent of all medications in the third world are not what they claim to be. Dilution is common. In most cases, even in the U.S., it's only a misdemeanor, seldom prosecuted, partly because it's hard to prove that's what killed a patient. WHO says counterfeit drugs are most common in China, India, Hong Kong, Japan, and in African countries. It says nothing about Internet drug marketers who sell products in the U.S., which they buy in China, India, Hong Kong, Japan, and Africa.[13] One study of anti-malaria drugs in Africa concluded that 50 percent are counterfeit.[14] In China, even veterinary medicines are often faked.[15]

Americans pay vastly higher prices for medicines than most other countries in the world. It is U.S. prices that drive the innovation process that generates new medicines. Maybe Americans assume that in our high-cost, high-regulation society, we are safer from counterfeit drugs. Eban says the opposite is true—that our higher prices have lured counterfeiters here. In 2003, there were more discovered incidents of drug counterfeiting in the U.S. than anyplace else in the world. (In 2004, we were edged out by Colombia.) We have a hugely powerful regulatory apparatus in the FDA. But it is not in the business of catching counterfeit drugs. People who are known to be counterfeiters go on making millions a year, again according to Eban. They pay fines. Chances are they never do jail time.

How big is the business of selling counterfeit medicines to sick people? If it were a corporation, it would be in the Fortune Top 20 companies in the United States. Think McDonald's and P&G put together. Think $75 billion a year by 2010.[16]

Now think about how bad it is

Procrit, or Epogen, is a cancer medicine. A weekly shot costs about $2,000. A pharmacist in Kansas City diluted doses of Epogen, and sold the extra, also diluted. Over 10 years, 4,000 patients were his victims.[17] Some of them would have died anyway, but nobody takes a $2,000-a-week injection unless there is a pretty good reason. How much moral distance do you put between a person who fakes a life-preserving medication and an outright murderer?

China has had numerous cases of babies who died from malnutrition, after being fed counterfeit milk powder.[18]

There is a large population of people on maintenance medications for high blood pressure, high cholesterol, for blood thinning and other conditions. In the U.S., by the time a person is age 50, there is a better than even chance that he is taking one of them every day. One in ten worldwide is phony.[19] There is no way to know the precise impact of faked or adulterated prescription drugs. Maybe the patient would have died anyway. Maybe he would have gotten well anyway.

It's not so easy to stop drug counterfeiting

Much of it begins with drug diversion. Stolen drugs are diluted, the bottles are refilled with cheaper ingredients, or they are relabeled to appear stronger than they are. It's not that tough to make a counterfeit label.

There are vultures who go off to the drug store with Medicaid patients, and buy the medicines prescribed for them, for money or for crack. Lots of drugs are stolen from hospitals. Lots of medicines are given to doctors as samples. Some find their way into the hands of wholesalers. Eban tells stories of drugs dispensed by Medicaid that were diverted and resold so many times the box wore out.[20]

The truth is, medicines that have been stolen or diverted must be considered no good. You cannot know if they were tampered with. You cannot know that they have been kept in conditions that preserve their quality.

However, it is also true that the prescription drug business has literally hundreds of small distributors. They come and go. They move from state to state. They are licensed, but maybe not inspected in any meaningful way. Diverted prescription drugs are not high on the list of things police departments worry about. If a small distributor gets caught, he would have to pay a fine for license violations. If he is caught again, he might even lose his license and have to re-apply. In the most drastic cases, he might move to another state.

It is also true that the price of individual medications goes up and down, frequently and by large amounts. They are sold at different prices in different channels of trade. Some, like hospitals, may pay half as much as others. This creates an incentive for an institution to overbuy and sell some of its discounted supply to a wholesaler. Whoever is at fault, diverted drugs with bastard origins re-enter the marketplace to be dispensed by reputable hospitals and big-name distributors.

Because U.S. consumers pay more for many medicines than people do in Canada or Mexico, there is pressure to "re-import" medicines, to buy them at the Mexican price or the Canadian price and bring them across the border. This is dumb, but politically controversial. States that don't have the money to keep their healthcare promises see this as a big source of savings. They're for it. The FDA is against it. The FDA says they cannot vouch for the safety of such medicines. That is certainly true. Counterfeits that make it into Canada or Mexico from the third world would certainly be among the medicines "re-imported." A very large Internet supplier from Canada has been caught shipping counterfeit Lipitor, Crestor, Zetia, Divan, Hizaar, Actonel, Nexium, Celebrex, and Arimidex to U.S. customers,. He was not arrested and has remained in the business under another name.[21]

There are many websites that offer to sell medications online at drastically reduced prices. This is only a bad idea if it is important for the medicine to be what it says it is. No major pharmaceutical manufacturer is selling its output to fly-by-night spammers for half the price that it charges Walgreen's.

How RFID attacks pharmaceutical counterfeiting: First steps

In some ways, the pharmaceutical marketplace is a good fit for RFID technology. It has high-value products, so a tag will be a small percentage of the price. It has big volume, so there will be economies of scale. It exists within the high-cost healthcare marketplace, where it may be possible to pass along the costs of protecting consumers from counterfeits. The severity of the problem and the embarrassment of two decades of abject failure by federal regulators may create some urgency.[22]

RFID tags can create true authentication: tamper-proof tags attached to tamper-proof packages. Bar codes cannot. The FDA has said that RFID tags are the best way to ensure that medicines are legitimate,[23] though it does not require them even in the most urgent situations.

Adoption has been slow. One study says 12 percent of pharmaceutical firms have some level of tagging in place, and another 3 percent have "widespread adoption."[24] It may be that others are waiting for standardized tactics and systems to be developed.

By 2007, ten major medications were tagged, at least in part. Among them were Viagra, Celebrex, and Oxycontin. Some are tagging cases and pallets only. But Purdue, for example, plans to tag every bottle of Oxycontin.[25] Cephalon produces Fentora, a widely-used cancer medicine which is also a Schedule 2 narcotic. It will be tagged at manufacture.

Drug manufacturers cannot solve this problem alone, even if they tag every bottle of every medicine. For RFID tracking to work, a product must be tracked all the way through the supply chain, from the factory to the pharmacist's shelf. So the participation of CVS, for example, which was an early explorer of RFID, and now leads the major drugstore chains as an implementer of RFID, is critical to counterfeit drug prevention.

Cardinal Health is one of the three giant pharmaceutical distributors, who together make up nearly 90 percent of dollar volume. Cardinal Health has begun to outfit its giant distribution center in California with an RFID system. California will soon require the tracking and tracing of pharmaceuticals. In fact, most of the progress here has come from state governments. There are laws requiring track and trace systems in Florida, Indiana, Nevada, and California.[26]

Seguro Popular, the giant of health insurance in Mexico has announced a requirement for RFID tags on medicines. The firm serves 20 percent of the population of Mexico. The Mexican Government will use the tag data to make reimbursement payments.[27] It is only fair to point out that, in this aspect of the fight against illegal drug trafficking, Mexico is well ahead of the United States.

All the way to ePedigree

State and federal laws say that medicines are supposed to go through the supply chain with a pedigree, papers that establish a chain of custody from manufacturing to final delivery. Every entity that takes ownership must be identified, and time of ownership described. Required fields include to/from addresses, transaction numbers, expiration dates and so on. It's called "paper" in the industry, and though this is not legal, it is possible for paper to be created far up the chain from the manufacturer. It is not unheard of for middlemen to buy medicines without any paper at all. Eban describes a marketplace full of "don't ask, don't tell" transactions, in which the price is more important than the documentation. Legal penalties are small and profits are large.

RFID can make the pharmaceutical pedigree a reality instead of a joke. A Tag-Data Security Infrastructure has been designed, with rules, specifications and protocols for tags and readers throughout the supply chain. An EPC number is programmed onto a tag. The product data portion is encrypted. But it is available as a pointer both for ePedigree and for any smart-shelf applications a drug store might want to use. Readers that have the verification key can authenticate the tag and decrypt the information.[28] Getting this or another set of standards approved should speed the process of ePedigree development.

Right now, we have incompatible state laws, an absence of universal standards, and an FDA paralyzed by the lawyers and lobbyists of small drug wholesalers, and by an adverse court decision those lawyers and lobbyists produced. Pharmaceutical producers who want to launch a system face large start-up costs, liability issues, and threats from professional privacy advocates. But all the problems are solvable once widely accepted standards exist. Every major drug manufacturer is looking for the next "blockbuster." Few of them will accomplish so much healing so quickly as they will when they harness RFID tags for the downfall of counterfeit drugs.

RFID and counterfeit brands

Two billion dollars worth of aircraft parts are fakes, in a typical year. Ten percent of high tech products are fakes. That's about $100 billion a year. The eight largest printer companies, which get much of their revenue from inks, lose about two billion dollars a year to counterfeits or substitutes. Add it all up, including counterfeit drugs and you're looking at roughly $450 billion a year. That's more than most whole countries produce.[29] One study claims that 8 percent of all trademarked products are counterfeit. The holograms used to prevent counterfeiting are now routinely counterfeited.[30]

Branded products face three kinds of look-alike competitors. Counterfeits pretend to be the brand. Generics are legal emulators of the brand, at a lower price-point and sometimes with lower quality. Substitutes are alternative competitive products, in situations where it doesn't cost much, or doesn't cost anything to switch from one brand to another. Manufacturers worry that they will face extra warranty or service costs from counterfeit or substitute components. Consumer advocates worry that RFID tags will be used to lock out legitimate generics and substitutes in situations where products must work together. For the consumer, counterfeits are bad, but imitations are good. It would be possible to design an office copier and its expensive toner cartridges with a simple tag-and-reader so that a substitute cartridge, without a tag, would keep the machine from working. This would be anti-competitive and would clearly damage the consumer. Existing laws address this problem, but it bears watching.

A maker of sophisticated medical test equipment has used RFID tags to lock out both counterfeit *and generic* versions of a necessary chemical reagent. The specifications of the chemical reagent are complex. Counterfeit and generic reagents that do not meet these specifications have entered the market. The manufacturer believes that these off-spec chemicals can degrade results, endanger patients, and damage the manufacturer's reputation. So the manufacturer puts tags on its authentic reagent. A reader in the machine has to see the tag on the reagent, or the machine won't work. Tags are serialized so that you cannot reuse a good tag with a fake reagent. This seems like a legitimate use of tags as an authentication device.[31]

Casinos have begun to use RFID tags to authenticate chips. A casino's chips are not that difficult to counterfeit, and tagging is much more robust than just switching the colors every now and then. They use multiple reader channels to ensure 100 percent reads. Casino technology is worth watching, as they are often innovators.

There is an emerging crime wave in the counterfeiting of high-end alcoholic beverages. When three-quarters of a liter of single-malt scotch sells for $150 or more, it's worth the effort to make a fake. Tags can be embedded in a glass bottle to authenticate its contents.[32]

A new smart label handles both identification and tamper-alert. It serves as a seal on a package. It can contain one kilobit of data. If the seal is broken the tag stops reading. Its printed antenna can bend and still function, unlike an etched antenna. The cost is down to ten cents in very large quantities.[33]

In aviation, a component part without a unique ID and a pedigree may not legally fly. In the future, RFID tags which are encrypted and cannot be rewritten will have the power to make this regulation real.

A consortium of designers has a task force working on business processes for RFID systems to authenticate fashion merchandise.[34]

Authenticating actions

The EU requires authentication of the processes in food production. A manufacturer must be able to prove *how* the product was made. RFID tags on a rack of cheeses can prove which racks went into which curing and ripening environments and precisely how long they spent there. They document the ripening process and prove compliance. A typical system may have sixteen different interrogation points. It records the product's identity and when it arrives at each stage.[35]

The Swedish Parliament uses RFID tags to authenticate votes. With a tag on the delegate and a reader at each place where an electronic vote may be cast, the system permits delegates to vote from anywhere in the chamber—not just from the delegate's own desk. People need not interrupt a discussion to cast a vote. The system also authenticates the presence of the actual voter, so it is not the desk that is voting, but the delegate.[36]

A Missouri school district requires school bus drivers to visit specific locations in the bus to check particular functions. An RFID tag time-stamps the driver's presence there.[37]

Some benefits of RFID authentication: a summary

By locking out counterfeits and below-spec imitators, RFID solutions can protect the quality and integrity of a brand. By keeping quality uniform, they protect customer relationships. They protect a brand from the warranty costs and support costs of inferior counterfeits. They protect the revenue stream on which the product's development was based. They lock out threats to customer safety. They can lock out refills and consumables that cause a product to fail in its performance promises. They permit investment in building and promoting a brand.[38]

The global problem of product safety

The key to product safety is reverse logistics. If you know what product comes from where, then you can respond to a problem with instant action and rifle-shot focus.

Reverse logistics can handle several recurring issues in product safety. One is defining the scope of the problem. If a quick service restaurant discovers e.coli in ground beef, knowing where that ground beef came from, and who else drew from the same source is critical. You can recall any product that might share the problem without declaring war on ground beef in general.

A related issued is how to fix the problem. If you can track a bag of contaminated spinach, or a bin of salmonella-bearing peppers, back to the field in which it was grown, that makes it easier to figure out how it went wrong, and how to keep it from going wrong again.

A third issue is knowing whom to hold responsible. Reverse logistics is the way to get that done. Commodity food products are good examples. It is not too difficult to know who manufactured a children's toy that turns out to have lead paint in it. It's much harder to know what to do if a bunch of people get sick in the same way at the same time. If there is microbiological contamination, the product may be gone before anybody shows any symptoms. We live in a global economy that is getting more global every day. Reverse logistics, by establishing responsibility for the safety of things sold, is the only defense against negligence by people on the other side of the world whom you will never meet.

The European Union has regulations that require a retailer to be able to trace food products all the way through the supply chain, and to trace everything it comes in contact with.[39] The EU is famous for making rules without much attention to the three issues of whether it is possible to comply with them, whether there is any relationship between the cost of compliance and the social value of the regulation, and whether there is the social will to enforce compliance. Nevertheless, such rules are being created at an accelerating pace. After awhile, regulations become a political tool in the competition between companies. Rules are used tactically, to whack a feared competitor, generally from outside the EU. It is in no way cynical to suggest that government protectionism will help drive the adoption of RFID food safety processes.

Reverse logistics are as important to producers as they are to consumers. Producers don't want to be held responsible for somebody else's mistake. Without a way to backtrack commodity food products, every producer is held hostage by his dumbest competitor.

Early identification

The best kind of food safety application is the kind that detects a problem before somebody eats it. Just about the only practical way to do this sort of thing is with RFID tags and sensor combinations, probably at retail.

There is now a sensor with a "nose" that can literally detect a bad apple.[40] Barrels everywhere breathe a sigh of relief.

A German meat-tracking project aims to protect the customer from meat that has spoiled in the supply chain. It gets at the problem from two directions. A semi-active tag with temperature sensors will document the continuous cold chain to identify and time any process problems. An RFID reader has an optical detector that analyzes the surface of the product itself. It measures the light spectrum in which chemical changes can be detected, and will see a surface degradation invisible to the naked eye, which signals that the product is past its best and is about to spoil.[41]

Animal tracking

Mad Cow is not the only thing to worry about. Important foreign animal diseases include West Nile Virus, Hoof-and-Mouth Disease, (eradicated in the U.S. but not worldwide), Equine Infectious Anemia, and *esicular stomatitis*. None is as threatening as Mad Cow, but one system would serve for all. Within the UK, individual beef cuts can be tracked all the way to the table, and cuts on sale at the supermarket can be associated with a particular animal.[42]

In Japan, this has been turned into a marketing advantage. In the Jusco Supermarket chain, individual cuts of meat have an ID number which can be associated, on labels at the meat case, with gender, slaughter date, test results, and whether the product is organically grown or not. They even post a picture of the smiling farmer. The Japanese culture is high in uncertainty avoidance, as Hofstede measures it, and, as is generally the case in such societies,[43] there is a willingness to pay for origin information on foods in general. As noted above, the U.S. examines one animal in 10,000 for BSE, but Japan examines every animal.

Since the Mad Cow incident in Washington State, Japanese imports of U.S. beef have been re-opened, but have never reached their former level, and have been shut-off when inspection errors occurred. Import restrictions are often political acts, especially in Japan. But Japanese journalists have argued for restrictions that are greater still, saying that U.S. practices are sloppy and dangerous and that the willingness of Japan to accept any U.S. beef at all is a political act, endangering the Japanese consumer.[44]

Korean consumers have marched in the streets to get their government to lock out U.S. beef.

Live animals or meat products might also have invisible attributes that could be carried on a tag. Tags could report vaccinations, special feeding regimens or organic processes, freedom from antibiotics or "good agricultural practices" which might add value.[45]

RFID tags could also be used with future identification systems that outperform the ear tag. Swift is developing a retinal scan for animal tracking. In Japan, Maple Leaf Farms sells a line of pork products that are identified by DNA. (A DNA test could conclusively associate an animal with a cut of meat.) Either version could create an ID that can be fastened on a meat product as an inexpensive passive tag.[46]

How RFID changes product recalls

RFID tags make food recalls faster. You can locate the specific batches of product you are looking for—on the shelf or in the cooler or in the backroom or in the warehouse—quickly, with the existing system of tags and readers, or with mobile hand-held readers. If all you have to rely on is batch numbers, it is not so simple. Batch numbers that are printed at the last minute onto an individual item label often are not on the outside of a case. Backrooms and distribution centers really don't want to have to open up cases of product that they are going to have to repackage and ship somewhere. The result is, a search may only find part of the batch. Items in the affected batch may turn up much later, long after the hunt has died down.

RFID tags can make food recalls more specific. In the U.S., 60 people each year die of *e. coli* poisoning, generally communicated through meat or produce. ConAgra faced a problem in 2002 in which some of its beef products were contaminated by *e. coli*. To protect the consumer and protect its brand, it recalled 18.6 million pounds of beef. One industry observer says it should have had to recall only 354,000 pounds of beef, but there was no way to tell which was which. Conagra did the right thing. They protected the consumer. Then they folded up their consumer beef business entirely.[47]

Tests by another meat producer, working with the USDA, developed a process for sub-lot sampling. A "lot" is, by definition, the smallest quantity for which firms keep records. It might be an individual animal in beef packing, or it might be a group of animals. It might be the output of one shift at a packing plant. In fresh produce, a lot might be the output of a field, or a group of fields, or one small part of a field. The bigger the lot, the bigger the difficulty in isolating the product that must be destroyed if a recall happens. Commodity producers, who have understood that every producer is at risk, are working with mock recall drills. This is an exercise that dramatizes and quantifies the advantage of small lot differentiation through RFID tagging.[48] A smaller lot size also tends to make the recall happen faster.

Food recalls often have a political dimension. Alar is a chemical once used as a cosmetic for apples. It makes them shinier and more regularly shaped. A rumor that Alar might cause cancer was magnified into a scare by national media. There was never scientific evidence that it did cause cancer, and eventually there was considerable evidence that it is harmless. But in the meantime, reporters quoting other reporters caused a nationwide recall and a lot of families lost their livelihoods. The removal of all apples that might have been treated with Alar produced a direct loss to growers of about $130 million. From time to time, the governments of Japan and South Korea run campaigns against imported foods. After the U.S. Alar scare, the government of South Korea detected traces of Alar in citrus juices imported from the United States. Alar is not used in citrus production. It was a clear case of apples and oranges. But juices were recalled and the damage to exporters was significant.[49] This is one of many examples. A tagging program can associate particular products with cultivation practices. There is no perfect defense against government misconduct, but tagging should discourage the traditional kind of attack.

The advantage of small lot sizes is not exclusive to agricultural commodities. In the first Iraq war, in the process of unloading Patriot Missiles from a ship, one missile got dropped and damaged. The crew lost track of which missile got damaged, so they had to ship all 20 back to the U.S., overhaul all of them and then ship them back to Iraq again. This cost taxpayers $21 million.[50] Remembering which missile they dropped would have saved a lot of money. An RFID system to track the missiles through the delivery process would have saved almost as much, and created an unalterable record. Today, the DOD is one of the biggest users of RFID tagging—perhaps *the* biggest.

You can tag a tire, to associate it with a particular vehicle *and* to associate it with a particular production line and shift. The debacle in which Firestone and Ford ended their generations-old alliance and blackened each other's reputation over a series of tire failures on SUVs might have been reduced or prevented by real-time, real-world data from a simple RFID application.

From time to time, tires are recalled, because of a known defect in manufacturing that makes them too dangerous to drive on. But most recalled tires are still on the road. Untagged, they are too hard to find.[51]

Bio-terrorism and food safety

The people who think about the unthinkable have started thinking about terrorism aimed at the food supply. Terrorism aims to generate fear, anarchy, hopelessness, or hate for the prevailing order. Food terrorism is the act or threat of contamination of some part of the food supply—prepared foods or plants and animals that become foods—probably with chemical or biological or radiation agents. Food terrorism was a frequent tactic in the Rhodesian revolution. The government tried to get poisoned foods into the hands of rebel sympathizers, with the hope that they would accidentally poison the rebels.

The first thing to understand about food terrorism is that there is no way to mount a perimeter defense. There is no way to protect all the sites where food is grown or processed. Companies can sharpen up their access control, but you can't protect a feedlot or an orchard or a grain elevator the way you can protect a factory. Packaged foods can be protected with tamper-proof and tamper-evident packaging.

The second is that food terrorism is more about information than it is about chemicals or poisons. The goal of a terrorist might be to make people feel threatened, damage some company's brand, cause a product to be recalled, or to become un-saleable. This could be done with a minimum of actual contamination and a maximum of threats and lies. The way you fight a campaign of small damage and big lies is with the truth, with immediate and accurate information. If food terrorism becomes another of the world's problems, some kind of RFID tracking will almost certainly be the solution. Just like a recall based on unintentional contamination, the defense is precise reverse logistics. Tracking information based on small lot sizes with specific points of origin is the way to reduce the impact of attacks on the food supply. RFID tags can also be used in some applications to detect chemicals with sensors, or to detect tampering with a pallet or package, and note when and where it happened.

The third and most important thing to understand is that the thinking of terrorists cannot be predicted. The most dramatic case of food terrorism in the U.S. was probably the man who introduced poisoned Tylenol into a drugstore's inventory. He bought four bottles, (before the era of tamper-proof packaging), poisoned them all them, returned three to the shelf, and used the fourth to try to kill his wife.[52]

Authentication and the marketing of invisible attributes

More and more, the marketers of agricultural commodities are differentiating their products through information about how they are processed. Food products may be described as organically grown, free range, free of antibiotics, free of genetic engineering, produced by a family farm, produced in a specific region, or a particular varietal version of a basic fruit or vegetable. Others define themselves as shade grown, water decaffeinated, produced without child labor, fair-traded, or as paying a satisfactory wage to those involved in production.

The use of invisible attributes can be a powerful marketing tactic. It justifies premium pricing. Two problems must be solved to make this tactic work.

One is how to *authenticate* a process claim. There are USDA standards a product must meet to be called "organic." There are third party authenticators, like Veri-Pure and Food Alliance, who are in the business of establishing standards for a particular claim, and reviewing a grower's processes to see if they meet the standards.[53] There are retailers, like Whole Foods, who play a similar role. Some retailers want a third-party audit of "Good Agricultural Practices." Both kinds of authenticators want tracking as part of the field-to-table process. Their credibility is all they have to sell, and they want to make certain that the product that bears their endorsement has come from the places they have certified. RFID tagging is the most credible way to provide this evidence.

The other is how to *display* a process claim. You can put a seal on the product label, but it won't be very big. There are about a dozen environmental ones, and they all look alike. If you have more than one attribute, if it is at all complex, if you want to give details, or if you want to make it prominent, it will almost have to be on a smart poster or a product tag, read by a cell phone or PDA or by an RFID table at the point of sale.

Tracking grains and seeds

In the past, corn and wheat and similar products in the U.S. have lost their source-identity almost as soon as they left the farm. The quality control point is the grain elevator. Once grains are sampled and found to meet a minimum standard, they are commingled. The elevator cleans and dries them. Everybody tries for the largest possible scale to keep down handling costs.

This is changing to cope with genetically modified (GM) grains. The people who created GM grains thought they had created a spectacular cost savings with no change whatever in the end product. In the most common example, corn plants were modified to be resistant to a particular low-cost grass-herbicide. It is necessary to treat cornfields for grass control to raise a profitable crop, and this is complicated because corn itself is a grass. If a grower can put down a $12-per-acre grass product instead of a $50-per-acre product, food becomes cheaper and more profitable. The change in the plant itself is far smaller than the changes created by hybrid seed corn producers going all the way back to the 1940s. But the language "genetically modified" has turned out to be hugely unfortunate. Customers in risk-averse cultures, especially in Europe, want no part of it, and it must be segregated from other product, thereby giving back much of the savings. The defining disaster was an event in which GM corn was mingled with other corn, which was then seen as "contaminated." The people who used it as an ingredient had to take processed product off the grocery store shelves.

Now farmers have to segregate all the way back to the field. They must protect against cross-pollination, use different harvesting and transportation equipment, or hyper-clean them between shipments. Often they need certification by third party authenticators, and again, these authenticators want tracking all the way back to the field. Container tagging with RFID is a robust and economical way to get this done.

Tracking fresh produce

Fresh produce in the U.S. is about a $25 billion business—half sold at retail and half through foodservice. In many cases, the shipper markets the product on behalf of the grower, and deducts a fee.

Almost all the market conditions that encourage RFID tracking are present in the produce business. Unlike grains, produce comes in a small container, all from a single supplier, so there is one entity to harvest the benefits of tracking. Some of the critical quality attributes such as size and grade, have to be identified early, after picking and before shipping. Significant information has to accompany the product. This, too, encourages tracking. Produce is a fragile product transported for long distances. If there is a problem at the end of the line, those involved will want to know who is responsible. This too encourages tracking. To minimize recalls, growers want to associate a product with a particular part of the field, another reason for tagging. A lot of produce wants to be labeled organic. Some may want some sort of certification of "Good Agricultural Practices" as well. Third-party certifiers for such processes want products they certify to be tracked.[54]

The best solutions are combined solutions

Products get tracked in situations where the benefits most clearly outweigh the costs. One of the advantages of RFID is that data can be used by several entities for several reasons. Another is that it is available all the way from field to retail.[55]

The benefits of tracking are higher where markets are larger. They are higher for high value products. They are higher where there is a higher likelihood of something going wrong. They are higher where the cost of failure is higher. They are higher where a problem is fixable if it is identified. They are higher where the value added by intangible attributes is high. They are higher where value can be added by different people or companies co-operating with each other.[56]

Food safety adds powerful reasons to track products, both at the pallet level and at the individual package level. As the cost of tracking by RFID falls, and as people find out about more of the benefits, RFID tracking will continue to grow.

Admissions, Permissions, and Tickets

What is a ticket?

It's what you need to get on the bus. It's permission to enter. It's the purchased right to consume or take part in or use something. This does not capture the allure of a product that people may stay up all night to buy, the urgency of a product whose value may soar to irrational heights, or the curious rituals surrounding a product that law-abiding people routinely purchase illegally from scalpers on the street. Football season tickets at large U.S. universities are willed to favored children. Stadium Skybox admission has become a business gift for VIP customers, so valuable that it may not be openly offered to a politician unless public records are kept.

Many times, a ticket is membership in a group. There may be other privileges attached. The dessert cart at the White Sox Diamond Suites is an event to be savored and atoned for. But the essence of the experience is membership in an elite group—a membership that, even though it lasts only a couple of hours, is valuable enough to be a memorable event. It might not have much to do with the event itself. Skyboxes and stadiums always contain lots of people who don't care very much about the game.

What about the person who commutes to work everyday on a train, bus, or light rail? That's a different kind of membership, but it's definitely an ongoing relationship.

Here's the point: *a ticket is a relationship product that is sold as if it were a transactional product.*

There are lots of exceptions, but that's the rule. It is the rule primarily because of the physical nature of tickets. If, instead of a thin slab of cardboard, you issue your ticket as an RFID tag, with memory, processing

166

power, and wireless communication, you can add all kinds of new functions. Most important, you can empower the relationship. Marketers by
the thousand are expensively seeking relationships with customers who
seek no relationship with them. Here, all you have to do is give wings to
the relationship that already exists. You don't have to pay extra for the
relationship functionality of RFID tickets. In fact, you get a bunch of
other benefits along with it.

What makes a good ticket?

Tickets should cut costs. At a White Sox game, employees at a turnstile
must handle each bar-coded ticket and orient it for a reader. RFID tickets,
on the other hand, get read by an electronic reader, which unlatches the
turnstile for one revolution, to let a patron in. After you buy the reader,
you pay only for electricity.

The cheapest kind of tickets for mass transit used to be paper tickets
with a magnetic stripe. Now, RFID tickets are cheaper still because they
cut the cost of maintenance. Magnetic stripe readers have metal contacts
that must be physically touched by the stripe on each ticket. They require
significant repair every time a would-be free rider sticks bubble gum in the
reader. The RFID reader is a sealed box. The difference in maintenance
alone provides enough savings to justify a switch. This study gets done over
and over, and the answers never change. RFID tickets get cheaper still as
manufacturers switch to an etched antenna rather than a coil.

Tickets need to be quick. This is hugely important in mass transit. People
do not want to wait at a turnstile for their tickets to be read while a train
arrives and departs. The London Oyster ticket to the Tube is read in less
than one-fifth of a second.

Tickets need to be convenient. Customers who take mass transit every
day have to think about whether they have their ticket. RFID tickets can
be downloaded to a cell phone. Wave the phone and walk on by. Tickets
in contactless "vicinity cards" can be designed to be read without being
taken out of the customer's purse.

Tickets need to be reliable. RFID tickets in mass transit operations deliver a near-100 percent read rate, an improvement over magnetic stripe tickets, which generate a fair amount of free rides. RFID tickets can be designed with a short read range to eliminate accidental unlocks.

Tickets need to be hard to counterfeit. An encrypted RFID tag is not particularly vulnerable to fakery. It is conceptually possible to make a fake tag, but most of the protection is in the system rather than the ticket.

The importance of this protection was underlined in the summer of 2008, when a pair of academic researchers published a procedure for hacking and then counterfeiting the basic MIFARE card, perhaps the most widely used ticket and access control device in the world. In the brutally Darwinian world of hackers and cryptographers, this is considered a normal process. What they did, under the protection of freedom-of-research laws, and while on salary from the taxpayers of Virginia, has created vulnerability, danger and loss for many thousands of public transit systems and building security systems all over the world. In some cases, soldiers were mobilized to replace the security which the distinguished hackers undid. But within a few days, systems designers were finding quick and inexpensive ways to put security back into the system whose security had been destroyed. Future RFID tickets will need to be built with more robust cryptographic protections. But note the speed with which the providers of readers, notably Feig, identified ways to restore and rescue MIFARE.[1]

Tickets should be revocable. A radio instruction to readers could turn off an RFID ticket.[2] If a customer loses his paper ticket to the ballgame, he is out of luck. If it were an RFID ticket, the vendor could shut it off and issue another one.

Tickets can empower collaboration. Amsterdam tourists can buy an RFID ticket that admits them to 25 different museums, plus public transportation. It can be programmed to match the length of their stay. All of the museums cut their ticketing costs, and quite a few of them get extra traffic by sharing.[3]

The relationship factor

Those are the extras. The key advantage of RFID is the ability to empower the relationship between the ticket-holding customer and the service provider.

A relationship ticket should be flexible. Think about mass transit. RFID tickets can contain all kinds of combinations: frequent-rider discounts to reward loyal customers, or pay-by-distance plans that encourage short rides, or bring-a-friend plans that encourage excursions. That's hard to do with paper tickets, but easy with RFID tickets, which can download a complex set of package-deal instructions with a few keystrokes, at a kiosk or from a PC. When a customer's RFID ticket goes into his cell phone it can download instructions on the road.

A relationship ticket can permit add-on purchases. Transit tickets could become a payment device at parking lots by the station, at fast food places, and convenience stores. This is an important potential source of revenue. The U.S. market for mass transit is about $8.6 billion. The market for Quick Service Restaurants is about $131 billion.[4] A small slice of revenue from that larger market might heal some financial ills in mass transit.

A relationship ticket can encourage loyalty. An RFID ticket could collect data for loyalty programs. No one has repealed the law that says the best source of business growth is additional purchases from your current users. It will be argued that collecting data about a person's presence or absence is a serious privacy violation, but it need never be done without the purchaser's approval. A customer might be willing to spill the beans about the fact that he takes the train from the suburbs to downtown and back five days a week in exchange for some free rides on weekends.

RFID admissions for transit

Mass transit is the biggest early user of RFID ticketing. About 50 percent of all the contactless payments in the world are mass transit tickets. Major systems include Hong Kong, Seoul, Singapore, Shanghai, London, Paris,

Berlin, and Chicago. The Chinese Railway system is switching to RFID transit cards. It ordered 125 million single-use RFID tickets for the first year of the Shenzen-Guangzhou commute. The national system has about 3 billion commuter-trips annually. Adoptions on this scale can change the cost of RFID tickets for everyone.[5]

Cost cutting

Besides the savings on maintenance costs, there are savings from eliminating fraud and counterfeit tickets. A major German system says it lost 25 percent of revenues due to fare dodging in the magnetic stripe era. They solved it by switching to RFID. The Moscow Metro had an enormous counterfeit problem. It gives maintenance costs and counterfeiting as the two key reasons for switching to RFID tickets.[6]

RFID tickets may be stored-value cards, with many rides paid for in advance. This improves cash flow. Or they may aggregate to monthly bills, which cuts transaction costs.

RFID tickets can eliminate the costs of processing cash. Many transit systems have kept a cash-or-ticket process on board each vehicle. This is expensive and creates cash management problems. Most RFID systems do not have cash in the vehicle. A rider can buy a ticket with cash or card at every stop.

Huge cost savings happen when RFID tickets morph into the rider's cell phone. Some pre-phone systems spend 16 percent of revenues creating and dispensing tickets. That shrinks with a switch to RFID, and shrinks far more when tickets become virtual—data existing inside a cell phone, replenished by phone or online. Systems in Germany, Hong Kong, Japan, and Korea have cell phone tag capability. One German mass transit system has used Nokia phones with integrated tickets for years.[7] Tickets in phones never have to be printed or dispensed.

With RFID, mass transit systems can outsource all their revenue collection costs to a credit card company. MasterCard has a tap-and-go card for New York subways. Not all governments want to outsource, because it eliminates patronage jobs, but some will find the savings irresistible.

The biggest source of potential savings is also the hardest to quantify. It is a truism in mass transit that improvements in the customer experience create an increase in ridership, which can change the economics of a system. Mass marketing has generally failed to increase ridership. Cost increases have driven ridership down, but every system has a pool of non-users for whom mass transit might make sense. Improve the experience and bring them aboard.

Improving the customer experience

RFID tickets create an immediately noticeable difference in the process of getting on the train, subway, or bus. Where there is a line of people, tiny changes in the amount of time it takes each person to get his ticket verified make a big difference in speed. An RFID ticket doesn't need to be lined up and slid through a reader, or pushed into a slot and retrieved. It takes milliseconds to be read. Wave and walk through. The line disappears.

A ticket on the Paris Metro can be carried in a customer's cell phone, and renewed online. It still works even if his phone battery is dead. And he can move it to his next phone if he changes phones.

A stored value ticket doesn't make the passenger find out each time what the cost of the fare is. Research suggests that an RFID card significantly reduces the *perceived* cost of the trip.[8]

RFID tickets can remain usable when fares change.

There are several ways to change the ticket's "form factor" which might improve the experience. An RFID ticket could be a hard card carried in the customer's wallet. It can be in her wristwatch. It can be in her cell phone, or in a shell around her cell phone if she has an old phone. Or a city could use all of those systems and let the customer decide.

There are cities, like Hong Kong and San Francisco, with several separate mass transit systems. RFID tickets can improve the customer experience with a single stored-value ticket that combines all of them. Some groups of systems are using a "combi card," with a magnetic strip *and* an RFID tag, while they unify their technology.

But the most important experience improvement grows out of real-time visibility. It is possible to get beyond aggregate data—to understand how individual riders use different parts of the system. You can know moment-by-moment how demand changes. With this clearer picture, there will be opportunities to manage traffic better. Service changes, which cause great pain for individual riders, can be made with the benefit of a better understanding, and emergency response gets better.

Empowering the relationship

Most everything that improves the customer experience will improve the relationship. But if you think specifically about making the relationship work better, more possibilities come to mind.

With RFID tickets, you see what's happening. You can experiment with group discount programs and measure participation. The Moscow subway has successfully targeted students, and can tell exactly how big a discount will optimize the system's revenue. Lots of systems offer a student discount. Before RFID, few could tell if they had picked the best price to maximize student ridership, without leaving money on the table.

German systems that use tickets embedded in Nokia phones can check riders on and off the system, so they can charge by the distance traveled. They can offer a smaller price for a shorter ride, and maybe pick up extra short-trip customers. There is a cultural variable here; some societies would see this price-by-distance as nit-picking, but to Germans, it feels fair.

Airports and air travel

This is an evolving area, where security needs and convenience needs are in deep conflict. The experience of commercial air travel has never been particularly pleasant, but surely it is horrible today as never before. Passengers are summoned to arrive two hours early for a one-hour flight. They stand in long, slow security lines, dutifully taking off and putting on shoes, getting out and putting away computers. They deal with frequently changing rules. The most hardened road warrior stands in line not knowing if the stuff that was permissible last time is permissible this time. The rules in Denver are not the rules in Des Moines. Newspapers carry stories about people denied the right to fly because their name is the same as that of a potential terrorist.

There is not a clear relationship between airport hassle and airport security. The climate has moved from a sort of patriotic co-operation to something like compliance-under-threat. There is every chance that it will deteriorate further. When a flight attendant stops passenger boarding and calls security because someone from a previous flight has left a cell phone on board, not everyone on the now-delayed flight is grateful for her vigilance.

Although there hasn't been much visible innovation here, there are certainly opportunities. Here's one. The Dubai Airport uses contactless smart cards to eliminate lines for some travelers. A 100k frequent flyer can register in advance and get a card with a biometric ID. They've been issued to customers from 35 countries.[9] The card automatically opens a door to the departure gates. A security person watches to see that uncarded people don't go through the gate. For the population included in the program, it provides a higher level of security than our domestic lines. But the politically incorrect will note that it is now quicker and easier and *much* more pleasant to board a flight to New York from the Middle East than from Chicago. More to the point, some smart marketer has created a relationship ticket that enormously empowers the Dubai Airport's most important customers. U.S. marketers take note.

When a passenger who has checked baggage onto a flight, does not board the flight in time for departure, the airline has a problem. It seems imprudent to let the flight take off with a suitcase sent by someone who has decided not to go along, but it's so time consuming to get that bag back off the plane that a flight could miss its departure window. RFID tags on each checked bag make it easier to find one and get it off quickly.

The U.S. is exploring a track-and-trace system that would link passengers with both checked and carry-on baggage, from the point when a reservation is made to the exit at the destination airport.[10] This can only be done with RFID technology. It could have been done five years ago. It would be far more meaningful than the current charade.

Digital network access

Admission to the IT system of a business or institution is an increasingly urgent problem. Systems that bar the door with nothing more than a password and username do not lock out anyone who really wants to get in.

However, the answer is not to make admission harder for the authorized employee or customer. An employee, client, consultant, or supplier who expects to use a single laptop in many different locations in the workplace, and wants instant access to printers, projectors, large-scale display devices, or other services, is expecting admission to get easier, not harder.

An RFID tag is a quick way to invoke a distributed wireless operation. The tag can be an easy-to-carry, powerfully encrypted, nearly unbreakable ticket that provides instant access to a network or to nearby wirelessly available, digital appliances. *It configures a link without user expertise.* It is quicker than typing in a username and password, and vastly more secure. And if lost, it can be disabled remotely, instantly.

Business people who make laptop-driven presentations at customer offices will find it a great mercy, as it frees them from learning each customer's network eccentricities.

RFID tags and/or readers in cell phones let the user launch an instant peer-to-peer network simply by touching the peripheral device with the phone.[11] It's easier than Bluetooth and doesn't require a separate piece of equipment.

The tag can also secure the laptop itself. Link the individual's RFID door-opener access card with the laptop's RFID tag, and suddenly, you cannot go through certain doors carrying somebody else's laptop.[12]

Buildings and facilities access

A pioneer use of RFID, the "prox card" is probably the most common admission device for offices, warehouses, and factories. It is quicker and easier than a mechanical key, and harder to copy. It can be turned off remotely, which solves the HR problem of getting a separated employee to return keys. It can admit people and record who has entered, or at least whose card has entered.

Its inherent flaw is the inability to know who came in when one employee holds the door open for another. However, there are ways to address this. For cost reasons, prox cards are usually low frequency RFID tags. If a UHF card is used, it can be given a read range of up to 25 feet so it reads the employee's admission card when she goes through the door even if someone else opens the door for her.

In buildings where high security is worth maintaining, you can add an extra layer with a digital facial recognition system. In this application, a camera picks up faces coming up to the door and matches them to a template in a database, or matches them to a template in their own admission card. It could be designed to alert Security if it doesn't get a match. It could also track assets carried in and out.[13]

Employees of Homeland Security have a single card that gives them access to the office, access to the computer, access to the wireless network, and a ticket to the subway. It contains a fingerprint record, and the card reader also reads fingerprints. You could argue that they didn't need to protect all these things. Or you could argue that they got all the controls beyond access control for free.[14]

Business systems with a card for access control can be used for physical access, access to the data system, time clock and attendance, cashless vending, process control, POS terminals, cafeteria payment, and automatic admission to the parking garage.[15]

Parking admission can be automated so that a tag in the car opens the garage or parking lot gate without the need to show a ticket. Tag systems can be designed to count the cars and turn on-or-off a "This level full" sign.[16]

Finkenzeller points out that you could make a prox card for cats.[17] A home can be fitted with catflaps that stay closed, but respond to the tag on your cat's collar. This plugs an energy leak in cold climates, and it will not admit the unauthorized; cats open the door for nobody.

RFID toll collection on highways was discussed in Chapter 5.

Argentina has a stored-value card for bridge tolls. A tag permits the motorist to pay in advance for a specific number of crossings. Here, the convenience of the card is assumed to be worth a toll that is higher than the cash toll at the same bridge. It fits in a marketplace where many people do not have credit cards, and where people driving on business need a way to shift the cost to their employers.[18]

Edinburgh has an RFID solution for express buses. RFID tags notify traffic signals when a bus is approaching an intersection. The traffic lights adjust their timing to give the bus priority and move it through the city faster.[19]

Entertainment and sports marketing

Amusement parks and theme parks will often charge a large amount for admission, but not charge for individual rides. In the pre-RFID world, this solved a lot of problems. It reduced the amount of cash that must be handled, although food concessions still needed individual transactions. It cut labor costs by eliminating ticket-takers at rides. It eliminated situations where small children had to handle cash or have a parent with them to handle cash.

But it also has some expensive problems. First, to cover the costs of "all you can ride," the park confronts the customer with a high and visible upfront price. This is the single biggest barrier to getting more frequent visits from the park's best customers. Second, it may seem absurd to non-riding caregivers. Grandma and grandpa want to take the grandkids to the park, but they don't plan to ride many rides. They look at the total upfront cost and see a very big number. Do you make grandma stay home? Do you simply make fewer visits to the park?

An amusement park always has a small number of rides with high demand—newer, wilder, or cooler rides. Some of these are a great deal more expensive to build or to operate. The one-price system doesn't let the park manage demand by charging more for these. It can only manage demand with discouragingly long lines. And lines are a well-known negative for the whole category. There are ways to palliate the problem, but there ought to be a way to make it disappear.

As you might imagine, RFID provides a solution for each of these problems.

A customer wearing a bracelet can "count" the rides taken, and be charged for actual use, with a much smaller admission charge. The bracelet requires no ticket takers. It recognizes the difference between grandma and the kids she brought to the park. It permits pricing to manage the demand for super-rides. It can be used to eliminate the necessity for cash at food concession stands and at game machines. A printed summary of rides taken can be delivered as the family exits the park. Most important, it does not make the family confront a three-digit ticket price at the point when they are deciding whether or not to make the trip.

Shift now from the rollercoaster to the golf course. Here a major barrier to sales is the need to schedule a tee-time in advance. It may not be possible to make a last-minute decision to play golf. Out-of-towners or golfers who don't have a foursome may feel locked out. This can reduce the load factor in courses whose traffic is uneven. It creates a situation where tee-times are not taken, and yet potential customers go unserved. Long term, it makes courses create less capacity than they could potentially fill.

The Wentworth Club in Surrey, England, has three golf courses. Two use traditional tee-time reservations. But each day, one course is available for individual members. They can simply go to the first tee and tee off after those in front have reached the second tee. A tag on the golfer's bag monitors use, and excludes non-members. This harvests a chunk of business that would otherwise be lost. The same "ticket" is used to monitor and charge for tennis courts and spa visits.[20]

The 2006 World Cup in Germany used RFID tag tickets. They reduce counterfeiting and fraud, which is a huge problem for paper tickets at an event where demand is non-rational. They can be used to provide all kinds of packages in which people buy precisely the access they want without buying a complex package of individual event tickets. A ticket can be shut off if a spectator must be removed from proximity to the games because of hooliganism. RFID tickets deliver real-time sales information about concessions or auxiliary products, so that capacity can be quickly moved where it is needed most.[21]

As in mass transit ticketing, when you can embed a ticket in a customer's cell phone, you can make the process yet more powerful. Make a reservation by phone or online, and have the ticket downloaded to a cell phone. The customer probably won't lose that ticket, and he can make it an open-ended ticket, in which auxiliary charges are added at the event, and billed to his credit card.

Discover Card has proposed a way to order multiple tickets and transfer some of them to other people's cell phones.[22] This fits nicely with new behaviors created by cell phone use, in which people going to an event together arrive separately and meet at the venue using phones to find each other.

Limited-access beaches are using RFID wristbands with a 10-foot read range. Managers walking by can tell whether people are ticketed without bothering them.[23]

Season tickets and relationship strategies

A season-ticketholder is a self-designated key customer. He accounts for a big share of revenue. In the NBA, suite, club, and season-ticket holders are 60 percent of tickets sold, but more than 80 percent of a team's ticket revenue.[24] Furthermore, season-ticket holders are probably the most growable source of revenue. They are committed. They have resources. And they will be spending lots of time with you.

Season-ticket holders can be assumed to have a positive perception of the relationship: it is why they are spending so much money. The marketer starts with lots of "knowns" about them. You know where they will be at a particular time. You know some purchase history. You know some friends and family. Most important, the relationship takes place in a context of high emotional involvement. Nobody jumps up and down and yells at the rent-a-car place.

Given all this, *when you shift from a ticket as a transaction device, to a relationship device that empowers some two-way communication, you create a uniquely powerful marketing situation.* Following are some ways to make the most of that situation. Some cannot be done without RFID functionality. Some merely piggyback on the relationship that an RFID ticket helps to create. They grow directly out of the CRM strategies in Chapter 3.

Loyalty building

When a ticket is an RFID tag, it's easy to offer special access. Take a lesson from airlines and casinos, and offer some kind of entrance—a special set of turnstiles, a special door—that's just for season-ticket holders. That provides both quick access (which is a huge perk in a stadium environment), and a form of recognition. Lesser mortals can see that your customer is a season-ticket holder. Special parking is already part of most season-ticket packages. It can be made automatic, and much more efficient than with paper passes if an RFID card opens the gate.

An RFID ticket becomes a loyalty card, valuable in all the ways that a casino card or a supermarket card is valuable. Loyalty-card marketers keep finding little things to give away to the valued customer. (In the sports venue, it's easy to catch them coming in.) The loyalty card can be the basis of point promotions, that add bonuses as the customer adds purchases. When tags are in customers' cell phones, it will be possible to see those points in real time, and see the positive consequences of one more hotdog for everybody.

Additional sales

Who's a better prospect for an upgrade to more expensive seats? The season-ticket holder knows all the advantages and disadvantages of his particular seats, and has spent time thinking about what his options are. At one franchise that's expanding to fill a very big stadium, marketers have mooted a promotion in which season-ticket holders get offered one-time use of a luxury box, or of super-premium front-row seats. Hey, sampling builds sales. In a real-time RFID environment, the team could even offer special seats that are unoccupied as a last-minute upgrade, the way airlines do.

Oddly enough, season-ticket holders are also good prospects for *additional* seats. They are logical hosts, bringing friends to a venue where they understand and value the experience. Whatever can be done to seat guests near their hosts, and to give them a chance to get together before or after or at breaks in the action will help to make this work.

There's an opportunity to make the most of unused seats. eBay's Stub-Hub ticket re-selling service and its competitors have created a market with *$3 billion* in revenues. They are marking up and re-selling the team's own product. Some of this revenue can be recaptured with an electronic product that makes it more convenient to resell tickets through the team itself. Flash Tickets, an electronic precursor to RFID, has begun to offer such a mechanism. But it's an early hybrid, in which tickets have to be printed out on the spot at the stadium. That means delays and line-standing. An RFID system can simply clone a one-time ticket from the original. Now it's more convenient to resell through the team.

Food and beverage sales get a boost from the loyalty-card structure. Already in some stadiums, RFID tickets are contactless payment cards that speed up concession sales. (More on this in Chapter 11.)

The market for souvenirs and regalia is transformed by the loyalty-card environment. It gives the team the ability to personalize—to sell, for example, a jersey that is autographed *to you*. It gives the season-ticket customer the ability to buy some other fan an individualized gift that others could not buy. Ask your kid how much that's worth.

Information products might be the perks that pay back the loyal customer. Media guides, score cards, insider newsletters, and the like can both acknowledge the relationship and make the fan an expert to his friends. A personally addressed message from the coach or the manager to the individual season-ticket holder is a powerful statement of relationship, but it doesn't cost much.

For the marketer with imagination, the big opportunity is the creation of additional experiences, perhaps to be earned with loyalty points. The unique emotional climate of the sporting event creates value in experiences that are not operationally difficult or expensive to deliver. Throw out the first pitch. Be the one who shouts "play ball." Get a pre-game photo with the pitcher. Get special close-up access at practice. Send a "tell-it-to-the coach" message that actually gets acknowledged. Get invited to an off-season meeting. Get access to pre-game party facilities. Get an event catered with the stadium's unique food and beverage products. There's a lot the marketer can do to create a memorable moment that rewards the loyalty of the season-ticket holder and brings him back next year.

Working with outside sponsors

RFID tickets change the marketing situation for a team's sponsors and partners. A sponsor can get a much clearer view of his audience, of precisely who he is promoting to. A franchise can solicit voluntary information sharing: help us help the team.

Coupons or dealer-visit promotions or any kind of trial-building offer can be a lot more powerful where a little audience information exists—where the sponsor knows what kind of car you drive or whether you're a heavy business traveler, or some other qualifier.

Promotional marketers worship "engagement"—whatever it takes to create an enthusiastic interaction. The sport situation empowers engagement. Games within the game, trivia contests, if-we-win-then-you-win promotions, answer-the-question-to-enter-the-sweepstakes promotions, and so on, all get their power from the event.

Finally, the brand itself can be the provider of a special experience. The ultimate premium is an event that recognizes your special interest, your favorite player or something like that, and makes a memory around it.

There are certainly limitations. Sports marketers can learn from the big casino groups, which oversee all the communications aimed at their key customers from all their different brands. They make sure all of them are appropriate to the brand, and they make sure that there is not an avalanche of offers. They protect the privacy of the customer, while making it easy for the customer to share information.

A ticket is a relationship. RFID makes it work.

CHAPTER 11

RFID in Payment Systems

Payment systems change behavior

The big-picture theorists of marketing talk about brand, positioning, engagement, and relationship. Retailers and service providers who are nose-to-nose with the customer think a little bit more about payment systems. It matters whether you pay with cash or check or credit card or debit card or automatic withdrawal from a bank account. It changes what you buy and where you buy it and how loyal you are. To understand how RFID is transforming this part of customer behavior, look at three effects of payment systems.

Quick and painless credit creates an enormous lift in spending. A vending machine that gets upgraded to use plastic gets an instant 20 percent sales increase, and some products in that machine (ones that need lots of change) will get a 50 percent bump.[1] In supermarkets, people start spending about 20 percent more per trip as soon as they are freed from the limits imposed by cash-in-pocket. This is true whether they are actually borrowing, or just moving money from a checking account with a debit card.

Speed of payment is a critical part of the retail experience. It can be a reason to prefer. Debit-card commercials dramatize the horror of being caught behind some cave-person who actually wants to write a check. The system for bar-code scanning your own purchases at a supermarket or home center is cumbersome, fallible, and annoying. But it's also popular, because everyone fears being stuck behind the person at the Home Depot who is checking out with all the parts for a two bedroom house. People will stand around at a Starbucks while their latte is constructed from scratch, and then be visibly impatient with the time it takes to pay. Retail studies always say the most annoying part of the store's experience is waiting in line.

Keeping *records* of payment by a particular customer makes loyalty more important, more visible at the point of sale. But it also makes things take longer. Now every trip to the bookstore or the grocery store or the drug store means fishing out a bar-coded loyalty tag, and adding a step to the transaction.

Historic shares are shifting

Cash used to have three advantages: seller preference, universality, and anonymity. The anonymity part still works. Cash is what people use to buy things they don't want others to know they are buying. However, in a society obsessed with data collection, this advantage creates a disadvantage. To get a measure of this, go out to the airport and try to buy a ticket on an airplane for cash. One way. Departing today. They may eventually sell it to you, but you will go through a hassle that you wouldn't believe, and you might end up on some databases where you do not want to be.

Seller preference and universality are waning fast. AT&T cell-phone stores warn customers in advance that stores "are not equipped" to handle payments made with mere money. If you don't have plastic, don't come in. The bus driver won't take cash. Citibank has bank branches that won't handle cash. Only if a customer is already carrying plastic will the bank handle cash, via its ATM machine. To make a deposit, customers must have the bank's own proprietary plastic. When a bank won't take money from strangers, you know cash is not universal any more.

The new advantage of cash is risk management. If a person carries cash, all he can lose is the cash he is carrying. But every time he uses a mag-stripe debit card or credit card, he gives someone an ill-protected opening to steal his financial identity, or empty his checking account. Cash may be the safest way to pay. Anyway, cash is rapidly losing share of transactions to debit cards.[2]

The personal check is also in decline, especially at the point of sale. It is slow, and not everyone takes checks. Checks were 50 percent of non-cash payments in 1995. Today they are less than a third and falling fast.[3]

The credit card companies built their business from the top down. By the mid-1990s, they had captured all the customers who are good risks for lending. So they shifted their focus to debit cards, which pay by deducting from a checking or savings account. Originally debit cards were aimed at low-income households. But they have come to be seen as faster and more convenient than checks, and now hold a broad market. By 2009, they will have replaced one-fourth of cash payments, and the vast majority of payments by check.[4]

If you have your paycheck "direct deposited" to your checking account, that's an Automated Clearing House (ACH) transaction. If you tell the gas company to deduct your monthly bill from your checking account, that's an ACH transaction. Service providers love them—they are the cheapest possible way to collect. For credit and debit cards, they are a fearsome competitor.

RFID creates a new high-speed segment

A contactless smart card permits tap-and-go instant payment.

Informal research at a neighborhood supermarket suggests that a conventional magnetic stripe (magstripe) credit card, with a signature on a screen, creates an actual transaction time of a little under two minutes, including getting out the card, getting it oriented and slid through the reader, getting the transaction approved, and signing the little plastic screen. This does not seem like an eternity, except to the people behind you in line. But there is a whole set of situations that people avoid because of the transaction time required by a magstripe card.

Consider the football game. A football game is a three-hour series of pauses and delays interrupted by just over seven minutes of actual play.[5] Yet, if you go to a football game, you will find that it is difficult to get yourself to the food stand, acquire food, and get back to the place you came from without missing a chunk of the events you paid to witness. No matter how many food stands there are and how well organized they are, a football stadium puts a big bunch of people in a small space, and they cannot be attended to quickly enough. Time spent suppresses purchase.

An early experiment in Seattle's Seahawk Stadium showed that using a contactless smart card produced an 18 percent increase in concession sales.[6] It requires a very small capital investment and it vastly increases customer satisfaction. If the transaction is small enough (currently under $25, but this will probably move up), then you don't need a signature. Tap and go. Faster than debit cards. Faster than cash.

A 1,600-seat theater sells drinks at a very healthy mark-up during a 15-minute intermission. Staffers moving through the crowd have mobile terminals. With contactless smart cards, they harvest every potential sale. They can serve and collect from everyone who wishes to buy.[7] Until contactless cards were used, it was never possible to know what actual demand was.

Contactless cards shift transactions, especially small transactions, from cash to plastic. In a MasterCard study, customers who acquired a contactless card used the card 18 percent more than they had used its magstripe predecessor. About 75 percent of those new transactions were less than $25.[8]

The Octopus, a contactless smart card launched in 1997 in Hong Kong, was originally a ticket on the Hong Kong ferryboat system. Merchants around the ferry docks wanted to take advantage of its convenience, and were allowed to join the system. Hong Kong has a strong self-service culture. It has crowd problems at convenience retailers. By 2002, Octopus had 160 merchants and 25 percent of its transactions were unrelated to transit. Now ATM cards, local credit cards, and Nokia phones in Hong Kong routinely add Octopus functionality. It is estimated that 95 percent of the "economically active" population is using Octopus.[9] Lots of them never get on the boat.

Gas stations seem like a perfect situation for speed-based payment. It is an interruption in a process where speed is important. It is completely impersonal. Time spent there feels like time wasted. In bad weather, there's an even more powerful incentive to hurry. Mobil introduced an early contactless technology, the SpeedPass, in 1997. It got an enthusiastic response from a small segment of the market even though it requires carrying a separate piece of pocket cargo that doesn't fit in a wallet, and that can only be used at a few places. Sales lift is one extra tank per month, presumably from greater loyalty.[10] Gasoline brands and service stations are hard to differentiate, except by location, so anything that creates loyalty is important. The first station to offer tap-and-go pumps will have a customer-experience differentiator, especially in cold climates.

How contactless payment works

The contactless smart card is a tag. The retailer's point-of-sale (POS) device is a reader. The reader powers the card. They exchange security information, and conduct a transaction.

As the reader powers the card, it transmits a random number. The chip performs an algorithm on that number using a proprietary code. It sends back the result of that computation plus a serial number. The host computer verifies this, verifies the amount and debits the credit card number.[11] The exchange takes between 150 and 300 milliseconds (count to one). The POS machine has an icon to signal that it accepts contactless cards, and a light that blinks to show that the exchange has taken place. "Tap-and-go" is not marketing moonshine. A tap is time enough to complete the transaction.

The rise of drive-through applications may create a niche for a slightly longer range card. On a message going the short distance from card to reader, powerful encryption and the use of random numbers that are different for each transaction mean that, even though it would be theoretically possible to eavesdrop, it would not be possible to crack the process. The real problem to be solved with drive-through applications is keeping the range short enough that you don't get an inadvertent transaction from a card that's still in the driver's pocket. The contactless system must also have an "anti-tear" process to prevent problems when a card has been removed before it is read.

If you watch people use a conventional magstripe card, you will notice that it takes a little bit of fumbling to orient the card and get it in the right position to slide through a slot. Many people require more than one try to get the stripe side of the card next to the stripe-reading side of the slot. This sounds insanely trivial, but, like the clicking protocols on an e-commerce website, the brief process of orienting the card becomes annoying, after you get used to tap-and-go. You can hold it any which way, tap the reader and put it back. The problem for contactless card marketers is that this only becomes important after you've switched to tap-and-go.

Where a larger amount or a higher risk is involved, the merchant might want a touch-and-confirm process, in which the user inputs a quick authorizer like a pin number. This prevents use of the card by an unauthorized person. (Magstripe cards at service stations ask for your zip code for the same reason.) It adds only a few seconds to the process, and can replace the tedious and useless signature.

Implementation issues

American Express, MasterCard, Visa, and Discover Card all have contactless products up and running. There are both credit card and debit card versions. The step-up to contactless is made much easier for them by the use of Combo Cards, which have both a magstripe and a contactless chip. If the customer pays at a retailer who doesn't have a contactless reader yet, the card still works. It is also possible to create a different kind of Combo Card, by putting a second chip on it, for multiple applications. An application in South Korea combines a contactless application for paying transit fares, and a magstrip that is used for a pre-loaded e-purse card.

A Javelin Corp. study counted 17 million contactless credit or debit cards by mid-2006.[12] They were used at 35,000 locations, skewed toward convenience. Early use was at quick service restaurants, convenience stores, pharmacies, theaters, and sports venues. In this study, 13 percent of customers had some experience with the cards. Their reactions were largely positive. A significant chunk of the early adopters had already tried Speed-

Pass. Trial is critical, and fundamentally changes consumer attitudes.

Of those not willing to try the product, 61 percent said fears about security were the reason why.[13] It just feels less secure to tap a card than to put the card inside a POS device. Older customers and low-income customers have been the most reluctant to try the new technology. Many people are still unaware of its existence. It is also true that the presence of contactless readers at retail is invisible, until you have a contactless card. The contactless icon on a point of sale terminal is less than half an inch high. You don't see it unless you are looking for it.

The real hurdle for implementation is the installation of POS devices with contactless readers at all the retail outlets that have only magstripe devices today. This is the classic dilemma in the credit card business: You have to get the retail side up and running so that it will be worthwhile for the customer to carry your card. Right now, 20 percent of retailers say that some or all of their locations can use contactless cards.[14] Some card service providers are helping to pay for the new readers, to gain retailer loyalty and get to the higher level of payments that comes with contactless technology.[15] Some POS terminals can be easily retrofitted with contactless readers.[16]

When a phone becomes a credit card

Japan and South Korea have tested and widely adopted systems where a cell phone handset contains a contactless credit card. Samsung and Visa did their initial South Korean launch in 1998. Several others entered in 2003.[17]

Nokia showed off a new U.S. cell phone with payment functionality at U.S. computer shows in 2007. MasterCard and Nokia have a large-scale trial run in the United States. Visa and Discover have each done several tests. Tapping a cell phone against a reader makes the indicated payment transaction. And you don't need to carry a separate card. Industry observers have been predicting this for years. A recent study suggested that half of handsets will have payment functions by 2010.[18] In Visa's tests, roughly nine out of ten preferred a phone-based system over either a fob or a card.[19]

One good thing about phones is that most people already know how to use them. Most users are comfortable browsing among multiple functions, and moving from function to function quickly. Test markets and user experience research say that a contactless payment system in a cell phone has a very fast learning curve. People are pleased with its speed and convenience. It is somehow more intuitive than the card alone. People may worry about where to touch the reader, but a single trial teaches them. A phone can use vibration or text or a light flash to signify that a transaction went through.

For larger transactions, phones can be used to quickly input a PIN code, instead of the slow and insecure signature.

Another advantage of phones is that they have their own power source. This makes a lot of useful features possible. A system might combine ID, access control, and payment, for example.

When a consumer's phone contains his credit card, a lot of changes become possible. First, it is possible, and not too complicated, to combine payment functions with other kinds of RFID downloads in the same phone.

Visa and Nokia held focus groups about a series of alternatives at Atlanta Arena, working with season-ticket holders. Along with the payment functionality, they offered downloadable player clips, updates, schedules, and so on. There is a long tradition of "affinity cards" which work off an emotional connection with a team, a school, or a charity. This is a much stronger affinity product, with sight, sound, and motion. It might be a reason to prefer.[20]

A phone marketed in Korea combines an NFC contactless payment system with "smart poster" functionality that lets a point-of-sale poster add a long-copy message retrievable over the phone. It is a pairing of two pieces of RFID shopping functionality.[21]

Discover and Motorola have tested an M-Wallet, a package of financial services. A customer can use it to make payments, check balances, and redeem electronic coupons. She can use it to make person-to-person money transfers, which is difficult with conventional cards. Paying the babysitter has been a cash-only transaction in the past, but perhaps no longer. The platform could support 50 types of credit card.[22] This is beyond a payments system; it is more or less mobile banking, and it is enough of a giant step that it will need some exploration to decide which functions add the most customer value.

A lot of credit card competition is based on the idea that a person has limited "wallet space" and can only make room for one or two credit cards. But he could carry more than one in a cell phone. This changes the competitive battle.

It may become very important for a card to be the "default" card in a customer's phone—the one that is used unless he selects a different card. This might dial up the competition among loyalty programs.

Here's another scenario. Remember the Automated Clearing House (ACH), where a service provider gets permission to deduct from a customer's checking account certain pre-approved transactions automatically? The company that gets to auto-deduct doesn't have to pay the plastic people their 2 percent. In the past, consumers have preferred pay-everywhere cards to a credential that they can use only to pay at Target or Nordstrom. But if a customer can carry 50 cards in a cell phone, maybe she would make room for closed-loop ACH systems that work for a single store or a small group of stores. With 2 percent of revenues to play with, the store with ACH payments could provide some pretty fancy incentives.

If a phone can carry lots of different credit cards, then maybe the cell phone manufacturer or the network service provider becomes a desirable alliance. Either or both could limit or veto the presence of particular cards—either general credit cards or closed loop store cards. Credit card products have always faced a chicken-or-egg problem with getting both customers and retailers aboard. Maybe now it is a four-part problem.

Maybe a cell phone becomes an ATM card as well. Phone-to-kiosk communication is available. Cell phone software would pull up account balances, credit limits, and interest rates and propose shifts either to minimize interest payments or to maximize available credit. It could also take into account incentive programs. Shifts from one account to another could be made by phone, and the software could list the relevant numbers. Discover Card has tested a phone-based product that lets the user review account balances and transfer funds between different accounts.

If the cell phone is part of a customer's payment system, it's inconvenient when he loses his phone. Most of us know someone who goes through a lot of phones. DoCoMo has a phone with built-in biometrics, so only the owner can use his phone.[23] Another often-proposed system would use voice recognition, so certain operations can only be performed with the owner's voice.

RFID and the single-environment payments system

Places where it is difficult to carry cash are switching to RFID-based payment systems that are quick, fumble-free, and precise. SnoMountain Skiers in the Pocono Mountains issues RFID bracelets that skiers can use to rent their equipment, access lifts, and pay for food at concession stands. It's an e-purse format, and it's easier than trying to get at an inside pocket with gloves on. It's a tap-and-go transaction at every place where money can be spent. Money can be loaded onto the tag at a kiosk, and unspent money can be recovered at the kiosk. Customers like it.[24]

Water parks offer a re-usable bracelet RFID system that combines keys with payment systems. It unlocks the customer's locker or hotel room door. Kids who wear a bracelet have access to cash, but don't lose it. There is a touch-screen application parents can use to find where their children are. Total spending per customer goes up where the system is adopted.[25]

What matters most to the customer

- Contactless cards mean faster checkout.
- They are easier to use than any other payment method. For small transactions, they require no signature, no PIN. There is no need to find change or count change.

- There is a choice of physical forms: cards, phones, fobs, watches.

- The card makes instantly available more than the amount of cash a person is likely to carry.

- The card can extend credit.

- Paying by card gives a customer a record of even low-value transactions.

- Low-value transactions can be switched to a mileage-points producing credit card.

- No account details need be revealed to a third party. Because the customer does not hand her card to another person, she has much better security than a regular credit card. In addition, there are new authorization and encryption safeguards.

- Users say the card is fun to use.

Speed is by far the most important advantage. Research results are consistent. It's faster than cash. It's faster than checks. It's faster than signature cards. In small transactions, it's faster than PIN debit cards. In just about every test, speed is the most compelling benefit.[26] Speed measurements for various payment mechanisms at various retailers are not very consistent from one study to the next. It may be that they are measuring different sections of the process. One of the ways in which contactless payment is different is that it eliminates the "fumble factor"—the process of orienting a card so it can be slid through a slot. If you look at the whole process of getting out cash or getting out a card, the time difference is greater. But even if you restrict the payment process to the handing over of a credential and getting the system to recognize it and send you on your way, contactless is faster. In one comparison, with a transaction under $25, paying with cash took 34 seconds. Paying with a magstripe debit card took 24 seconds. Paying with a contactless card took less than 15 seconds.[27]

American Express says its contactless card is 63 percent faster than cash and 53 percent faster than a magstripe card with no signature requirement. The simplicity of the process may make it feel faster than it is. Among payers in its study, 87 percent said it was better than cash, and 82 percent said it was better than conventional cards. A test in drive-thrus showed a 90 second saving in each transaction.[28] That's especially significant if all the people in front of you also have a contactless card.

Card marketers are assuming that contactless products will get most of their use in such high-speed, low-value transactions as fast food, movies, service stations, and the like. This segment was more than $165 billion in 2002 and was 95 percent cash.[29] Drugstores and supermarkets are also logical early adopters.

Some see convenience as more important. Customers who have tried the contactless experience note that you don't need to orient the card, or let go of the card. Customers exposed to the phone concept express relief at its simplicity, and at the idea of one instrument replacing a stack of different cards.[30]

The cell phone form is seen as superior. In one test run by Visa, 89 percent of customers preferred the cell phone over a card, a fob or a watch.[31]

Small transactions are better handled by contactless cards. The lower limit may be lower than marketers imagined.[32] Asked how small a transaction they would make with a card, one sample responded this way:

Under $5	22%
Under $3	12%
$1	19%
Under $1	22%

Interestingly, 85% would go over $50.

Security is seen as the key problem. People who don't want to try contactless cards cite security as a problem. Again, the cell phone may address some user worries about security. To the provider, the contactless technology seems much more secure than conventional cards. It has encryption and tamper sensing. The code for a transaction that is actually transmitted changes every time. Providers would offer freedom from liability for fraudulent transactions made with a small payments card.[33] New versions of the fob form can be locked to prevent unauthorized transactions. A SIM card can be "locked" or it can be invalidated if the phone is lost. It may be that security perceptions can be improved by marketing. The most powerful claim would be that contactless cards prevent the identity theft that is so easy with conventional cards. Card providers may be reluctant to make this claim.

Personalized services are *not* seen as important by customers. If you know what a customer buys, you know a lot about what a customer values. There are thousands of ways in which a merchant could use purchase data to make special offers to especially good customers, or to people at a particular point in their lives. Different products, customized, unique products, volume discounts, seasonal products, trade-up products, accessory products for things you already own. The list is endless. Conventional payment cards have learned to distribute slightly segmented coupons at the point of sale, often via the register tape. These could be personalized much better with data from smaller transactions.

It is important to marketers and card marketers to understand that this opportunity to personalize is seen as a negative by many, probably most, customers. Relationship marketers have proven over and over again, in many different businesses over many years time, that relationships and personalized service can be made attractive. But they have also learned not to go too far on the first date. After the data exists, very small steps, with lots of reinforcement, can be made toward improving what the seller provides to the buyer who is better understood.

What matters most to the merchant

It's all about lift. Time after time, advances in payment systems have produced large, predictable, consistent increases in the amount spent by a customer on each visit. It is a case that no longer needs to be made. Merchants enter the discussion optimistic about new spending generated by new payment systems. In a MasterCard study, contactless cards increased transactions by 12 percent. In an American Express test market, contactless card transactions were 20 to 30 percent larger than cash transactions.[34]

The speed benefit of contactless cards is seen as *enabling more transactions at a given point*, not just in short-window situations like the football game described above, but in supermarkets as well. Mass merchants work hard to reduce the wait at checkout, and spend a lot of money to do so. A system that hands back a bunch of free seconds with every customer will change the operational arithmetic in many ways, not least of which is labor cost.

Protection from fraud and non-payment. In the two-sided business of payment systems, almost all the cost of non-payment is borne by the merchant side. As card providers became harder to differentiate, they have competed more and more by improving the detection of fraud risk.

One key source of risk is the use of credit cards for Internet payments. Internet credit card payments have ten times the fraud rate of payments at brick-and-mortar locations. It is a "card not present" transaction, and the merchant assumes all of the risk. Think of this as two separate problems. First, it is difficult to know online that the person who is giving you a card number is actually who he says he is. This is the "authentication" problem. Second, if a customer sees a transaction on his monthly statement that he did not make, the customer can repudiate that transaction by notifying the credit card company. In a "card not present" transaction, a repudiated transaction is charged back to the merchant. If the payer was present when the charge was made, the charge cannot automatically be repudiated. The payer must get the card issuer's approval to repudiate the transaction, and this is not automatically available.

Contactless smart card systems can be designed to address both problems. The American Express Blue Card can authenticate the card holder online, if the user's computer has a card reader, which more and more laptops do. *It empowers non-repudiation, and turns a card-not-present transac-*

tion into a card present transaction.[35] Thus, it addresses both authentication and charge-backs.

Improvements in customer loyalty. Increased satisfaction due to faster checkout and simpler payment is easy to detect,[36] but card marketers believe that this will be only one of the drivers of improved customer loyalty. As contactless payments replace cash, the merchant gets a greater ability to understand purchase patterns and product preferences. This should help to build a relationship in several ways. First, it can empower more personalized and higher-value services. Second, it supplies the information necessary for personalized promotions. It would be possible at a service station to present an individualized promotional offer while gas is being pumped into the car, for example.[37]

Limitations, competitors and concerns

Several times, privacy lobbying organizations, or reporters with limited technical background have engineered a particular kind of stunt story. They bring a portable reader, generally with a large or highly specialized antenna, into close proximity with a contactless card, and demonstrate that the card can be read by an unauthorized person. They show that they can recover the card number and the expiration date, and suggest or imply that such information is equivalent to the power to spend someone else's money. Again and again, the card providers reply that the security is not at the card level; it is upstream. There is a whole system of authentication codes, and second-by-second changes in the transaction code and ways to make it evident if someone is trying to tamper. It is worth remembering in this context that the business model generally offers zero liability to the customer, and places liability with the card provider. This might be considered evidence that the providers who designed the system believe it is secure, or at least more secure than the credit and debit cards on which businesses depend today.

What is most real about the security problem is the sort of preconscious perception of a problem by users. The idea of wave-and-go payments has a funny feel to it, and this sort of perceptual problem has killed many other products. Card issuers are counting on successful experiences in a fairly slow roll-out to remove this concern.

There is also an issue in the combo cards, which card issuers are using to smooth out the start-up. Combo cards have both magstripe functionality and a chip for contactless transactions. The extra security advantages of contactless will not be available until new adopters replace their combo card with a contactless-only card, or with a phone-based card.

There are certainly limitations in the kinds of RFID technologies that can be used for payment systems. Tags must be powerful enough for sophisticated encryption. They must have an extremely short read range, and be durable enough to be carried around for years. They are nothing like the low-cost tags stuck on a pill bottle.

Most people who study the future of payment systems believe that the true future competitor will be some kind of biometric, where all you have to bring is your body. Tested and functional pay-by-fingerprint technology exists, and was adopted by several supermarket chains. The leading provider has ceased operations, but consumers like the technology, and it will be back. It is important to design biometrics so that there is no way that it profits someone to run off with the finger or eyeball of an unwilling other. This is probably solvable, but the problem can be avoided altogether with RFID technology.

Where RFID fits best

A few key applications will drive the adoption of contactless smart cards, and of smart cards carried in cell phones. Eventually, they may replace all the pay-with-plastic functions and many more.

The first step is face-to-face micropayments, taking business away from cash because cards are quicker and simpler and let you pay when you don't have cash with you. You can make a few office copies at a Kinko's, and pay with a wave. (Everybody at Kinko's is in a hurry.) You get a record for your expense report, and you get out the door without waiting to be waited on, which would take longer than it did to make the copies.

The second step is face-to-face larger payments wherever speed is an issue. A phone is easier to get out than the appropriate credit card. You don't have to insert it. You don't have to sign. At most, you punch a four digit PIN, which is quite quick on a cell phone.

Single-environment systems permitting face-to-face payment of larger amounts are already succeeding.

The third step is adding other RFID functions besides payment, installed in the cell phone: a mass-transit ticket, an ATM card, a reader for smart posters at retail, or a system to transfer funds from one account to another. Or, all of the above.

The fourth step comes from the marketer: personalized coupons or similar promotions that cut the cost of useless offers to unprofitable customers allow more impressive offers to identified and profitable customers.

The fifth step is the one from tactical loyalty systems like the one described above, to strategic loyalty-and-payment cards that focus more tightly on the needs of the merchant than the general-purpose credit cards do. At this point, several things become possible. Retailers might use the Automated Clearing Houses to give a store-specific loyalty card the power to deduct from a checking account. This would, as described above, save the whole credit card transaction fee for the merchant, which could fund some fairly amazing incentives.

Retailers with a single-chain loyalty and payments card could use transaction information to better understand customer differences, and personalize their communication with an existing profitable customer. Perhaps the money that is currently wasted on coupon-bearing mini-catalogs dumped by the sackful on Sunday newspaper subscribers could be used to craft individual messages that do not try to sell a toaster to someone who bought your toaster last week. Let the whole roaring torrent of indiscriminate mass-market promotions resolve itself into a tiny trickle of pure and personal gold.

CHAPTER 12

RFID and Patient Relationship Management

MEDTRONIC AND DIABETES

You can be a diabetic and still lead a happy and a healthy life, but boy, you have to work at it.

Being a diabetic means managing the level of glucose in your blood, twenty-four hours a day. Blood glucose changes constantly, in complex ways, affected by the interaction of diet and exercise and time of day and the amount and timing of insulin added to your system. *Most of the time you cannot know exactly what your blood glucose level is.* A few times a day, you prick your finger and get a measurement. A couple of times a day, you inject insulin to reduce it. You keep a journal of measurements and meals and exercise and injections and how you feel—and you strategize with your doctor about how much and how often to add insulin.

And if your management isn't exactly right, that's a problem. Un-metabolized glucose is a poison. It causes heart disease, blindness, nerve damage, kidney damage. There's always the danger of a hyper-glycemic event. The more poison, the more problems.

Now there's a new management tool. Medtronic has an application that combines *continuous glucose monitoring,* automatically collecting data from a blood glucose sensor under the skin—with *continuous response,* from a monitor and an insulin pump. They're linked by short range radio communication: tag and reader. And they're tiny enough to let you live your life. The transmitter is about the size of four quarters in a stack. The pump is no bigger than a cell phone.

It's a dramatic example of what happens when you replace a few scattered data points with a continuous flow of information. You can't do a fingerstick every 30 seconds, 24 hours a day, seven days a week. But now you can get as much information as if you had. Patients are learning that their blood glucose levels jump around faster and more widely than they imagined. Doctors are able to graph the data and associate it with events in the patient's journal. They can overlay data from one patient with data from another, and use that information to refine their treatment strategies. *Management gets better.* Changes happen—lots of them. Here's the biggest.

Knowing what's happening in real time, and responding more precisely means (in one study), you can reduce your average blood glucose level from about 7 percent to about 6 percent. Doesn't sound like much, but that reduction is associated with *a reduction in the side effects of diabetes—blindness, nerve damage, and so on—of 25 to 33 percent.*

That's a life-changer. A life-changer for diabetic women who want to have a child. Precise glucose management is necessary to prevent harm to the unborn child. A life-changer for people who have hypoglycemic events without noticing them. Or while they are asleep. A life-changer for diabetic children who must hold off the side effects for a long time.

There's a difference in personal comfort as well. Patients are still doing fingersticks to double-check blood level before re-setting the insulin pump, though that step may someday be removed. But instead of two-to-seven injections every day, patients experience one sensor insertion and one pump insertion every three days, (to minimize the risk of infection).

You can be a diabetic and still lead a healthy and a happy life. Now it's a little bit easier.[1]

The idea of patient relationship management

The American healthcare system sets a standard of professionalism for the world to follow. It is a network of practitioners with extraordinary skills, constantly consulting and advising one another, constantly updating their education, constantly scanning for new solutions, and finding them. It is supported by an engine of technological innovation, lavishly funded and unmatched in its torrent of new ideas.

Yet no service in our society offers such a horrible customer experience. Some of it is inevitable. The suffering created by injury, disease, pain and fear, the final struggle that postpones death cannot be overcome, only palliated.

Some of it is avoidable suffering, created by administrative choices. Dangerous choices. Wasteful choices. Abusive and disrespectful choices. Erratic and unmanaged choices. *Old-fashioned choices* that do not make use of new learning about how individual relationships can be managed to satisfy individual needs.

Best in the world in the sophistication of its therapeutic technology, but in the management of relationships, hardly the equal of a first-rate discount store—this is the American hospital. Experience it and try to deny the diagnosis.

Maybe some of it is an attitude problem. Hospitals do not have customers—they have nearly powerless patients, who cannot simply walk away if they are not respectfully treated. The great revolution in customer service never reached the healthcare industry. Hospitals have taken up the tools of mass marketing, but outside of the highly competitive maternity segment, it would be hard to argue that they have competed by making the experience better.

Maybe some of it is a visibility problem. Hospitals are necessarily decentralized, juggling competing priorities, executing complex and controversial decisions around the clock at high-speeds in scattered locations. Maybe the power to see what is happening across the institution in real time could improve some decisions, prevent some errors, reduce some costs. It's worth a look.

These are strong words. To justify them, there must be at least the possibility of startling improvements. Judge for yourself.

Drug error prevention

A *New York Times* story collects some almost unbelievable numbers, most of them from the prestigious U.S. Institutes of Medicine. Preventable medical errors in U.S. hospitals cause 770,000 "adverse events" per year. Among them are 98,000 preventable *deaths*—more, the *Times* reporter points out, than are caused by drunk drivers and breast cancer put together. Thirty percent of malpractice suits involve drug errors or related problems.[2] This is an emergency.

Nurses learn the "five rights": the right medication, in the right dosage, at the right time, to the right patient, via the right route. If you are bustling from person to person, with a heavy schedule of routine tasks interrupted by emergencies, if you are joining a process in progress that started on a previous shift, if the patients are all dressed alike, and you don't yet know them by sight, and some of the bottles are pretty much alike and much of the time you work in the dark, well, it may be possible to make a mistake.

Another study[3] says things are worse than that. It says that drug errors at healthcare institutions injure more than 1.5 million U.S. patients per year. This doesn't include the patient's own mistakes. One error per patient-day is the average. That is a chilling statistic. *If you are in a hospital for one day, chances are that one mistake was made in your medications.* Two days, two mistakes. Some of these are small errors such as missing the "right time" by an hour or two, but some of them are serious.

The study by the Institutes of Medicine found a range of error rates from 2.5 percent to 6.5 percent. The errors had an average cost of $2,257 per event. According to the study, a hospital with 20 admissions daily would face an annual charge of $708,100 due to adverse drug events.

The Institutes of Medicine says preventable medical errors cost $17 billion per year.[4]

An early attempt to solve this problem tested bar codes on medications, bar codes identifying the nurse, and bar codes identifying the patient. If the bar code on the patient's chart and the bar code on the medicine container were incompatible, the system issued a discreet warning to the nurse. The test had two outcomes. First, it showed an astonishing ability to reduce errors. They were cut by more than 86 percent in the tests. Second, it suggested that the problem may be bigger than imagined, because it detected errors that would not have had an identifiable impact. Lots of times, the impact of a mistake is either not that damaging or not detectable. In this one test of the bar code system, from 6.2 million doses administered, the application issued 192,000 warnings of a potential error, and 152,000 of those warnings prevented administration of a harmful medication.[5] Consider the possibility that the hospitals tested saw themselves as in control and among the best, before they consented to such research.

It should be repeated that RFID systems have significant advantages over bar codes. Consider a system in which there is an RFID tag on the patient, one on the nurse, and one on the medication package. They don't have to be oriented to be read. (It may not be ideal to twist a patient's wrist to line up the tag with the reader.) They work in the dark with perfect accuracy. Most important, they are automatic. The caregiver does not have to add the task of machine-reading the three tags: only the task of looking at the screen to make sure there is no error warning. A typical system has a tag in a wristband, on the medication and on the nurse. When read, it shows a picture of the patient on a screen. It checks with a database that has new medication orders added to it automatically. It reads "ID Confirmed" or "Access Denied."[6]

Another type of problem involves medications that have lost their effectiveness for one reason or another. Some medications are highly sensitive to environmental conditions such as heat or cold or light, which may make them unusable long before their expiration date. Some vaccines can lose their effectiveness if they are removed from the cold chain for even a short period of time. Some medications used for battling cancer in the late, life-and-death stages of treatment, become useless if they have not been kept at the proper temperature.

It would be a cruel failure indeed if a patient in a life-threatening situation is given a medicine that doesn't work any more. However, changes in the effectiveness of a medication can't be measured without using up the product rather than treating someone with it. No hospital has the power to monitor a medication from its manufacturer, through a supply chain that may involve four or five changes of ownership, to the point where it is administered to the patient. There is a powerful negative incentive for shippers and distributors to raise the alarm about a hiccup in the cold chain, yet hiccups happen.

DHL, the shipping company, has a product that can monitor the temperature and shelf life of products in transit. It is an RFID tag on a long paper strip. It has a temperature sensor and a tag with a chip that records the sensor readings. The end of the tag hangs outside the box, so it's easy to read. The tag carries data on the product, its serial number, shelf life, expiration date, temperature requirements, and data on how shelf life would be reduced by specific temperature-change events. The result is information from a neutral source, which cannot be altered by shipper, distributor, manufacturer, or receiver. It creates a valuable warning and inexpensively prevents a potentially lethal error.[7]

Procedure error prevention

In a recent 30 month period, in the state of Pennsylvania, there were 125 "wrong site surgeries"—a gentle phrase that means the doctor cut off the wrong thing. The study identified 250 more "close calls."[8] A SurgiChip application uses a tag that sticks onto the relevant body part. The whole system, plus maintenance for one whole hospital costs about one-tenth as much as the average wrong-site lawsuit. Alvin Systems has another system that verifies "right patient, right site, right procedure."[9] The ID process can also be extended to taking blood, taking x-rays, performing diagnostic tests, giving a new mother the right baby, and transferring patients from one part of a hospital to another.[10]

A new application puts passive tags on surgical sponges and swabs, to alert the surgical team so sponges and swabs are not left inside the patient.[11] Sponges account for two-thirds of all the stuff inadvertently left in patients. Eighty-five percent of sponges left in patients are left after a manual count by a nurse says all sponges are out. The RFID application has a reader which counts removed sponges as they are dropped into a bucket.[12]

A product in trial matches patients with containers of transfused blood. It aims to reduce errors in which the wrong blood type is given to a patient.

Human error rates for blood transfusions have not improved over the past 20 years. A tag on the container is matched with a tag on the patient's wrist. It replaces a two-nurse, 27-step matching procedure with a one-nurse, 16-step matching procedure. The automatic nature of the process is expected to reduce errors.[13]

Another product aimed at the same application uses an electronic seal. The RFID tag locks the bag of blood to be transfused. The seal can be unlocked only when a handheld device communicates the identity of the correct patient. The lock will not open if the blood temperature is too high or too low. There is a unique ID for each element in the process, including the test-tube, the request form, the blood bag, the nurse, and the patient. It's quick and automatic *and its protections can't be skipped.*[14]

RFID tags can also be used to track specimens collected from a patient and sent to the hospital's lab, or an outside lab, for analysis. Hand-written labels created by busy individuals, writing down the names of people they have never met, pose a risk. One RFID solution comes from Smart Medical Technologies. The system includes RFID-enabled test tubes, vials, specimen collectors, and patient bracelets. The patient ID is linked to the specimen ID as the specimen is collected. Tags are designed to withstand refrigeration and centrifuge. They are encrypted. An extremely short read range adds another layer of security. The result is a system that can identify an individual specimen accurately and fast.[15]

Asset tracking

A healthcare technician in a modern hospital is a highly trained professional—an individual who can have a positive impact on the lives of thousands of patients. One study says hospital technicians spend as much as 40 percent of their on-duty hours searching for equipment.[16] What a waste.

RFID tags report the location of a wheelchair or respirator or other piece of moveable equipment moment by moment so it can be retrieved efficiently. Asset tracking in hospitals is one of the fastest growing RFID applications. Hospitals have lots of high-cost assets that must be moved from patient to patient, such as respirators, infusion pumps, pulse oximeters, defibrillators and hemofiltration machines. Hospitals are by necessity dispersed and decentralized management environments. Until now, there was no way central administrators could see where moveable assets are.

If movable assets are not easy to locate in a hurry, a secondary problem is created: the hoarding of assets by an individual department. It's not that hard to hide stuff in a hospital. If people put equipment aside, in a closet or above the ceiling tiles, so that it will be available when they need it, then it won't be efficiently shared and the hospital will need more pieces of that equipment than it would otherwise need.

There is also the problem of moving assets in an environment that is full of sources of infection, including hospital-nourished secondary infections. Sometimes it's important to know where things have been.

Hospitals measure their efficiency in part by the rate at which such moveable, sharable assets are used. The national average for moveable hospital equipment is 45 percent.[17] If hospitals could increase utilization, and buy fewer machines, the effect on healthcare costs would be significant. One study puts over-procurement at 20 to 30 percent.[18] Remember that these things are not cheap, and that most hospitals are trying to grow their asset pool to catch up with the most urgent needs.

Hospitals pay for assets by associating their use with a particular patient and billing for it. However, it's a decentralized environment, full of very busy people. Tracking usage for billing purposes is one of many tasks, and not the most urgent. A study of infusion pumps in a group of hospitals said the billing information was recorded 50 percent of the time. If all the usage was recorded and properly billed, according to this study, a 900-bed hospital would increase its revenues by $8 million annually.[19] This begins to make the cost of a $10 active RFID tag seem endurable.

Retailers refer to stuff that goes out the door without being paid for as "shrink". Hospitals have shrink too. Some of it is drugs. Most of it is equipment. An industry rule of thumb is $4,000 to $5,000 of shrink *per bed per year*.[20] Asset tagging may not perfectly eliminate shrink, but it should eliminate most of it and, before long, identify the sources of the remainder. Brigham and Women's Hospital in Boston had a recurring problem with the loss of EKG cables and cardiac pacers. RFID tagging revealed that bed makers were wrapping them up in sheets and sending them to the laundry, where they were discarded.[21]

One of the interesting things about asset tracking in a hospital is that many hospitals already have WiFi. That's the biggest part of the infrastructure for an RFID system. It can mean a large reduction in start-up cost. Another issue is how precisely you need to track a piece of equipment. Usually, you only need to know which room it is in, not where within the room. For very large pieces of equipment, you may only need to know its vicinity. This too reduces the cost of RFID asset tracking.[22]

Every manufacturer and systems consultant has success stories. Richmond Hospital claims a savings of 30 minutes per nurse-shift from not having to search for equipment. Miami Hospital started its RFID tracking project after discovering that it could not locate more than $4 million worth of equipment. Barrington Hospital says its inventory losses were immediately cut in half.[23] There are hundreds more.

Admissions, records, and information management

Emergency admissions. For as long as hospitals have existed, people have studied the process of getting life-saving care to an emergency patient as fast as possible. The best hospitals do this better today than it has ever been done before. Nobody thinks it's a solved problem. There are some places where RFID can help.

In cases where first responders treat patients outside the emergency room, or enroute to the emergency room, it's possible to tag, for example, fire victims or accident victims for tracking. When critical information is attached to the patient on a tag, before he or she enters the hospital, some repetition, some wasted time is eliminated. RFID provides an efficient way to do this. Input the patient's answers to questions on a PDA or mini laptop. Write that information to a wristband tag via a reader in the ambulance. Eliminate the need to ask the patient the same questions at the emergency room.[24]

An RFID tag for diabetics can speed diagnosis, especially if the patient is unconscious. There might not be much time to respond.[25]

More efficient admissions. If a customer goes into a car rental place where he has filled out all the records before, the rental company can pull up his data in a few seconds. If he comes into a hospital, where he has been before and filled out all the basic information, despite the fact that he is hurting, disoriented, and impatient, he will probably have to tell them many of the things he told them last time. It is the perfect opposite of a relationship: they will pretend not to know him, and he will have no good reason to go back to the same place next time.

Today an important chunk of the cost of healthcare is the admissions process. Technologically, it would be a slam-dunk to put a portion of your patient records on a smart card that you carry in your wallet. This could be used most anywhere, and could reduce the cost, delay, and anger that happens in the hospital admissions process. It would also permit automated copying of data from one system to another. Retyping data is the big source of errors in admissions.

There are huge issues with data privacy and data security in healthcare. Smart cards have good solutions for these issues. They can be encrypted. They can be PIN protected. They can be in the physical possession of the patient.

The process of admission *drives* the hospital's revenue cycle. Seventy percent of errors that lead to denied or "pending" health insurance claims are directly attributable to the admissions process.[26] Sometimes that means the hospital never gets its money. Sometimes it means that it has to get money from the patient that should have been paid by the insurer, damaging relationships in three directions, simultaneously. Almost all the time, it means the hospital gets paid later and spends more on getting paid than it ought to. A smart card system would automate the process. It would provide records data from someone who was not in pain or drugged up when the information was collected.

Sharing information efficiently. Health care in the United States is a team game. Lots of specialists are involved. Often the specialists are in different institutions, or in different departments of the same institution. Often different physicians see the same patient who comes in to deal with the same issue on different days. Often they take the same information from the same patient many times, manually. Smart cards could share information between different doctors treating the same patient. They could share information between different institutions treating the same patient. Not only would this deliver palpably better service to the individual patient, it would address one source of healthcare costs. As we enter a period in which the population is aging and likely to get a whole lot older, there will be more chronic illness. There will be a more diverse group of healthcare providers. There may be a more complex payments process. There will certainly be a more complex information flow. Contactless smart cards have the necessary privacy protection, data security, efficiency, and speed. They are overdue.

Medic Alert, whose stainless steel bracelet was a pioneer in patient data sharing, is switching to a smart card system. A card is a more efficient and detailed way of putting patient data at the point of service. It will speed retrieval of files, eliminate most of the repetitive form filling, and speed referrals. Medic Alert predicts that emergency first responders will carry cell phone or PDA tag readers.[27]

Patient access to patient information. It is a curious thing in our privacy-focused society that one person who cannot easily get access to a patient's data is the patient.[28] Often the patient cannot easily find out what information has been recorded about him, and has no way to correct or challenge information that might be wrong. Partly, this is a cultural problem, a symptom of the institutional disrespect of medical practitioners for their patients. A consumer has the legally protected right to look at his credit information, and correct it. His medical records belong to someone else. Partly it is a consequence of antiquated systems. There is solid, if unsurprising, research, from other service businesses, which says that if a customer has access to his own data, it will increase customer satisfaction. This access is hard to provide in the era before electronic medical records, which, oddly enough, we still are in.

Emptying beds earlier. As in other businesses, when you can see a process operating in real time, you can make improvements in it. When you can see where the bottlenecks are, you can redirect your efforts to go around them. Has a patient returned from a lab test? Instead of walking to the room, you can look at a screen and know instantly. Are there people waiting for a bed and people waiting to be discharged at the same time? If the caregiver knows that is happening, it might change the priorities on his or her endless list.

Nurses, respiratory therapists, and other providers operate with an implicit batch method. They have a list of things to do. They have limited ability to access information that might change their work list. Anyone who has ever stayed with a sick relative in a large hospital has seen it. The respiratory therapist comes around while the patient is off in x-ray. It's a wasted trip. The respiratory therapist comes around as the patient is being discharged. Another wasted trip.

It may take some time for better visibility to produce better processes, but the difference can be significant. A long-term test of RFID patient tracking produced some powerful results. Patients discharged by noon went from 20 to 40 percent. Incoming patients sent elsewhere dropped by 25 percent in critical care and by 60 percent in medical/surgical.[29] Hospital revenue is about throughput and throughput is about process and process is about patient visibility. Radianse says a 0.5 percent increase in throughput in a 300-bed hospital increases revenue by $3 million per year.[30]

A Houston hospital advertises a new level of service, called "concierge care." In this deluxe program, they pledge that a patient will be in a room within 30 minutes after entering the hospital, and that a nurse will be with the patient within three minutes after that.[31] An RFID patient tracking system makes it possible to keep such a promise.

A software platform from Aionex aims to increase satisfaction by recording patient preferences and offering a more personalized experience. It empowers patient self-service in some routine areas, and claims to make service delivery more consistent.[32] In a hotel, such a system would be routine. In a hospital it is revolutionary.

More efficient payments. A patient's smart card can identify the patient, define the relationship with the insurance provider, explain current insurance status, identify the primary care physician, and store data on healthcare activity.

Eligibility and coverage can be updated automatically, at the employer end or at the provider end. Electronic prescriptions can be stored on the card and eliminate some steps at the pharmacy. Data can be partitioned, so pharmacists can see the part they need, but not parts that must legally be protected. Information that would otherwise be unavailable becomes available immediately. Redundant tests can be eliminated. Calls to previous providers for records can be eliminated. The cubic volume of filed paper records can be reduced.

- For the provider, this means a quicker, more efficient, less expensive process.

- For the patient, it eliminates an annoyance at a vulnerable moment. It adds to the speed of response in an emergency. It may also improve the patient's access to the patient's records.

- For the employer, it provides a way to understand for the first time a major cost source in a timely way. Employers, aggregating de-personalized information, could renegotiate their HMO costs.

Such a program would require strict access controls. Access rights must be verified every time the card is read. This can be done automatically, and smart cards can meet all the HIPAA requirements.[33]

Storing the patient's blood, cells and tissue. LifeForce stores a patient's own white cells, frozen, at a secure location. The cells are identified to a particular individual, with the idea that they might be used later, in therapy for cancer or HIV or something else. RFID provides a way to catalog and locate these cells. It's more secure than paper records, and reduces the time and cost of locating them. LifeForce believes this model will be important in future applications where individual biological products are stored.[34]

Privacy issues are acute in healthcare applications

Some healthcare information is strongly socially charged. To know the private data on someone else's healthcare experience is about as deeply invasive of privacy as anything you can think of. The legislation referred to as HIPAA protects all individually identifiable healthcare information. It is[35] against the law to call a person by name in a doctor's waiting room. (Personal experience suggests that this rule is sometimes breached.) Under Garfinkel's RFID Bill of Rights, it might be necessary to notify the patient every time information is added to a tag-borne health record.[36] This is interesting, as it might be the only time a patient learns much of anything about what is on his health records.

Biosensors and telemedicine

The development of the technology of bio-sensors has been a series of explosions, so close together as to create a continuous roar of innovation. Idea after idea has emerged. Many situations and conditions that would be elusive even in face-to-face diagnosis, now appear to provide chemical, electrical, or optical signals of their presence, which can be captured by a sensor, stored on an RFID chip, and accessed by a reader. Maybe the day is coming when, instead of calling the doctor, the doctor will call you—having already identified a biochemical event, diagnosed the syndrome that produces it, and arrived at a prognosis that something must be done. Serious, senior practitioners are designing systems that will call the doctor before a patient has a heart attack and tell him that he is about to have one.[37]

Sensors can monitor everything from simple heart rate and bodily movement, to heart rate as affected by movement, blood pressure, breathing rate, pulse, several variants of blood glucose measurements, the presence of a great number of specific bacteria of interest, stomach acid in the esophagus, and glaucoma, among other things.

The more we learn about human health, the more sense it makes to adopt some sort of check-and-alert system that will identify certain vital signs for an individual, and notify that person or another person if those signs reach a dangerous level. The easy part of this is the technology: it's been possible to operate remote biosensors for a long time now. The hard part may be helping people decide whether it makes sense for them as individuals.

The existing applications look like Neanderthal predecessors of what will become the basics, but it makes sense to list some of them to suggest the possibilities.

- The basic "vital signs" with which a doctor starts a check-up can all be collected remotely, continuously, and relayed to whomever needs to know.

- You can monitor infant incubators remotely and continuously with sensors and RFID tags. If this technology can migrate to the home, perhaps it can provide a better defense against Sudden Infant Death Syndrome (SIDS).

- You can monitor heart activity, EKG activity, and the functioning of a pacemaker. Medtronic has a mechanism that can collect information from your pacemaker when you are at home, or when you are visiting another city, and respond to irregularities in a timely manner.[38] It is now possible for a cell phone to carry a heart monitor, in essence a reader for an RFID tag.

- Glaucoma is a group of eye diseases that lead eventually to blindness. After age 50, one in 200 suffers from glaucoma. After age 80, the figure is 1 in 10. A risk factor for glaucoma is high interocular pressure, and the most common treatment is drugs that lower interocular pressure. For people in early stages or at very high risk, there is a pressure sensor for the eyeball, attached to a very small RFID tag. An antenna in a pair of eyeglasses collects signals from the tag.[39]

- There is an RFID system for monitoring acid reflux. A passive tag detects stomach acid, gas, and water in the esophagus. Previous methods of acid reflux testing require a very uncomfortable procedure, using a wire through the nostril and into the esophagus. The RFID solution is, of course wireless. A patient wears a chip (1 cm by 1 cm) on the wall of his esophagus and a reader around his neck, underneath his shirt. A PDA in his pocket takes data from the reader. An endoscope removes the sensor when it is no longer needed. A digestible tag, easier still, is currently being tested.[40]

- Kodak has applied to patent an RFID tag that would be attached to a pill. The tag itself would be digested at the same rate as the pill. It would stop transmitting when the tag, and presumably the pill, have been digested. The device is designed to monitor whether patients are taking medicines. A rule of thumb: two-thirds of the time, people taking prescription medications actually follow the directions and take the right dose at the right time. There are applications where non-compliance is much higher, and applications where non-compliance is very dangerous. Seniors may have lower-than-average compliance, partly because they are taking several maintenance medications for problems whose symptoms are not immediately noticeable, like high blood pressure or high cholesterol. The tag would permit discreet intervention without checking up on people all the time.[41]

Tracking and monitoring patients

Real-Time Location Systems (RTLS) can help caregivers know exactly where within a hospital a patient is right now. There are several reasons why this is important. Patients are often sent from place to place within a hospital—to x-ray, to surgery, to recovery rooms, and secondary recovery rooms, and so on. In an operation where caregivers work in shifts, coordination isn't easy.

Some hospitals have visitation spaces away from the patient's room. There are advantages to this in two-patient rooms. More hospitals might consider separate visiting spaces if it were easier to know where a patient has gone.

Recovering patients are often asked to walk around the halls, for mild therapeutic exercise. Some hospitals, particularly those with longer-term patients, would like to give some patients the freedom to walk around within the hospital campus, but keeping track of the whereabouts of individual patients will burn a lot of man-hours. Nurses trying to give a scheduled medication can be more efficient if they know where patients are.

One hospital uses manpower to monitor patients with a high risk of disorientation from wandering. They use RFID tags to monitor low-risk patients. The result has been a 64 percent reduction in man-hours devoted to monitoring, which paid for the RFID system in its first year.[42]

Another hospital uses an RFID tag with an acceleration sensor, which can report that a patient has fallen.[43] Tags in development will use accelerometers to analyze a range of movements, but most are focused on falls.[44]

A psychiatric ward has adopted what it calls a Duress Card. The card has a button which can be pushed by the patient, when the patient is away from his bed. It gives an alert through the caregivers' monitor system. It shows the patient's location and photograph.[45]

Monitoring in-home care for seniors

The healthcare system is just beginning to confront an aging population.

Seniors will need different things from the healthcare system than were needed before. They will be more demanding than seniors have been in the past. They will not be as willing as earlier generations to make sacrifices in quality of life, for the convenience of the caregiver. Consider, for example, the increasing reluctance of a segment of the senior population to put up with "assisted living" environments. There has to be a way for people to get help as they get older without spending the last part of their lives in healthcare institutions. There has to be a way for people to get help, even as their own ability to evaluate situations and decide when to ask for help, gets weaker. There is research that says seniors will be willing to give up some privacy to be able to stay at home. In this context, RFID is a lot less intrusive than a camera, and probably a lot more powerful as well.[46]

Making clinical trials more efficient

New medicines in the U.S. are at the mercy of a painfully slow and extremely expensive testing process. Much of the $802 million spent by pharmaceutical companies to develop each individual new medicine, is spent on clinical trials. On average, they impose a 12-year delay on the availability of a new medicine.[47] People die while medicines that could help them are working their way through clinical trials. RFID could speed clinical trials, in four ways.

First is the remote administration and tracking of a large number of samples, all of which look as much alike as possible, but each of which must maintain a separate identity. Samples pass through the hands of several organizations. People handle them who don't have much information about the context of the trial. Trials are outsourced to a clinical research organization that may hand off tasks to several unrelated healthcare facilities. All this means it's easy to lose the identity of a sample. However, if the identity of a sample is lost, it becomes useless. If the identities of a significant number of samples are lost, then the whole test may be invalidated. *This happens.*[48] Samples get misplaced, misrouted, lost during transport, lost somewhere within the site of the healthcare facility, or lost somewhere within the site of the clinical research organization. Empty tubes sent to the investigator's site never get filled. Wrong tubes get returned. As a result, tests get cancelled and re-performed.

Sometimes it's necessary to track the environment around samples. Some medications are highly sensitive to temperature or light. If there is any failure to preserve the appropriate conditions, a change in results may cause a false appraisal of the drug's efficacy. Not only must temperature and other factors be sensed and monitored, the data must also be recorded and preserved for each sample.

It's also necessary to track the association of particular samples with particular patients, caregivers, and so on. As you know by now, all of these are easy tasks for RFID applications, but difficult and untrustworthy without RFID.

An EPCglobal study of clinical trials showed that, in an RFID-tracking environment, tests happened faster, start-up delays were reduced, trial errors were decreased, and the dropping out of test subjects was reduced. Real-time visibility of the process, and sharing of data among the participants led to modifications which are expected to improve the results of future tests.[49]

Reducing the cost and increasing the reliability of clinical trials is certainly important, but the time saved may be yet more important. If a promising medication becomes available even one year earlier due to the increased efficiency created by RFID, that is an important outcome. That is a change in the customer experience.

The United States is at a crossroads in health care. The current process is too expensive, too accident-prone, too inefficient, and too unresponsive to human needs to be tolerated. The solution to this will not be found only in fiddling with how health care gets paid for. It will be found in teaching that most inhuman of institutions, the American hospital, the value of patient-relationship management.

How RFID Works

This appendix is a short and simple explanation of automated data collection with radio frequency identification, focused more on concepts than on engineering. You might decide to read it before you've read the rest of the book, or cherry-pick sections that help you understand a particular chapter. If you want more detail, a list of very good books on RFID technology is at the end of this appendix.

Please note two things before you start. First, setting up RFID systems is still both art and science. None of these systems can honestly be described as plug-and-play, but nothing will be described here that can't be done. Second, RFID technology changes literally every day. This is a snapshot.

In a nutshell

This is how one basic RFID system works.

An RFID reader transmits radio signals at a pre-set radio frequency and interval. Signals go out hundreds of times per second. An RFID tag attached to an item has a built-in antenna. Whenever the tag is within range of the signals sent by the reader, it will pick up the signal and bounce it back. The tag will slightly change the signal to send back information. That signal, received by the reader, says "here I am" at a particular time and place.

This is how each component works.

Radio waves

Radio waves are electromagnetic energy. Passing an electric current through an antenna creates them. Radio waves travel through the air and pass through solid objects so any antenna within range can pick them up. They can be blocked or interfered with by metal, liquids, or by machinery that emits electromagnetic energy. Radio waves can transfer small amounts of energy—enough to power a computer chip for a moment or two. Radio waves can be modulated according to a pre-arranged code, so that they transfer information. A particular modulation communicates a particular piece of data.

Different antennas produce radio waves with different wavelengths: low frequency, high frequency, ultrahigh frequency, and microwaves. Different frequencies have different characteristics as message carriers. You have to choose the right one for the task and for the environment where the task is carried out.

Tags

An RFID tag is a radio transponder on a computer chip, with a small antenna alongside. It may also contain a small amount of computer processing power, and a very small amount of memory. In some cases, a sensor is attached as well. There is a wide variety of different tags, designed for different tasks. Tags may be contained in a paper label, a credit card, a plastic button, a glass capsule (for hostile environments), or a metal box. The plastic button tags generally have a hole in the middle for a fastener. Tags can be glued to, clamped to or embedded in a product or a product's package. They can be built into key heads, or inserted in machine tools. A tag inside a paper label can be stuck onto a package. There is also a way to inlay a tag into the corner of a cardboard box.

Power

Active tags have batteries that power their radio signal. Passive tags take their power from a signal sent by a reader. They bounce the signal back to the reader. Semi-passive tags have a small amount of power to drive a sensor or perform other functions, but still use the reader's radio signal to power communications.

The vast majority of RFID applications use passive tags. *Passive* tags are less expensive, currently under ten cents in lots of one million or more. But passive applications use *many* tags. In a typical passive system, tags are about 60 percent of the total cost. Tags on pallets or cases or items of retail merchandise are almost always passive tags.

Active tags have a broad range of prices based on their capabilities, but all are more expensive than passive tags. For tagging large containers or expensive assets, active tags are often a better fit. In those jobs you don't need so many tags and the tags stay around a long time. In such an application, most of the cost is in the readers, and active readers are cheaper. Active tags can have capabilities which passive tags do not. The battery in an active tag can power communications, processing, memory, and sensors, for example. Active tags can be designed with longer read-range than passive tags, or with the power to talk to other tags. Most sensor applications use active tags. There are tags that become active just for a moment, in response to a particular event, and then turn themselves off to save the battery.

Size

A typical passive tag has a chip about as big as the diameter of a pencil eraser, but paper thin. It also has an antenna, often glued or printed onto paper. However, chips can get much smaller. There are specialized tags as small as a grain of sand. Active tags can be many different sizes. They are often attached to big things, so their size is not important.

Read Range

Tags can be designed with different read ranges to fit different applications. A tag on top of a telephone pole might want a long read range so it can be queried from the ground. A tag on an ID card might want a very short read range—less than an inch—so only the reader it wants to talk to can detect it.

Storage

Different tags have different data storage capabilities. Some can store enough data for a volume of text or a short video.

The information stored on Read-Only tags cannot be changed. Read/Write tags permit the user to add information, or change existing information. They can be both read and written to from a distance. A third form of storage is called Write Once/Read Many (WORM). Information can be added to a WORM tag after it leaves the originator, but only once. After that, it's Read Only. That's useful when several firms are working together, but not all of them are permitted to change the data.

Most tags are used as a pointer to a cell in a database. The database contains all the information; the tag is just a license plate. Most system designers believe it is a better idea to keep the data in a database, and a mistake to send around a lot of data on the tag, just because you can. It's less expensive if you can do it that way. Larger memory requires more battery power.

Coupling

Coupling is the mechanism by which the tag and the reader affect each other.

Backscatter Coupling involves bouncing radio waves that were sent by the reader, back to the reader, from the tag. It's often used with passive tags in large retail applications. It can function at very short distances, or several meters away.

In *Inductive Coupling*, the reader's coil antenna generates a magnetic field. This field induces a current in the coil antenna of the tag when the tag gets close enough to the reader. The resistor on the tag switches on and off to create voltage changes in the reader, which communicate ones and zeros. It works only at short distances.

Sensors

Tags can carry many different kinds of sensors, to notice and communicate changes in temperature, acceleration, orientation, vibration, rpm, humidity, rainfall, wind, biochemical hazards, air pressure, acceleration, radioactivity, sound, light, toxicity, pH, gravity, blood glucose, and so on.

Sometimes sensors are combined with microactuators that can respond to changes in sensor readings by starting or stopping motors, or turning switches on or off. They can also respond to radio signals.

Sensors can also be connected to dynamic display devices such as electronic inks, holograms, flat screens, and so forth, to produce a message in response to a sensor reading or a radio signal.

Limitations

The effectiveness of tags can be limited by radio interference from machinery or metal or liquids, but users have found clever ways around most interference problems. Unless shielded, tags are affected by extreme temperatures and by moisture. The mass production of tags is still imperfect. In a large batch, some tags will be dead on arrival.

Designing a system and choosing the right tags requires an engineer. The engineer will think about read range, the kind of material or package to be labeled, the physical environment and what tag form and protection it might require, the standards and protocols you need to work with, and tag cost.

Prediction

Many people believe that current chip-built tags will be replaced by printed layered organic semiconductors, like a very small flat screen. If so, costs will fall. Some predict that printed tags will cost less than a cent. Others believe that a process called Surface Acoustical Wave technology will lead to a cheaper tag. Tag cost is the limiting factor in many applications, so it is the focus of extensive research and development activity.

Readers

The purpose of the reader is to interrogate tags and collect information about their presence or absence at a particular location. Readers can be stationary—at a doorway, loading dock, warehouse bay, or some other strategic location. They can be portable, as on a forklift truck. They can be handheld. (Note: WiFi can capture some data from tags without using a reader. This can be a cost-cutter in simple applications.)

Each time a reader sees a tag is one observation. (There may be hundreds per minute when a tag is in range.) An observation that differs from the previous observation (where a new tag appears or a previously observed tag goes away) is an event. The analysis of observations is event filtering. The event management system decides which events are sent on to middleware or an external application.

A single reader can collect data from about 800 tags per second. An "agile" reader is one that can interrogate several different kinds of tags.

Antenna design and placement are crucial. The goal is perfect coverage of a chosen area, and zero coverage outside that area. Many readers have separable antennas, to optimize coverage.

A reader is at the edge of a network, connecting the company's central Enterprise Resource Planning (ERP) system with this outpost in the real world. It is set up to handle *alerts, observations,* and *commands.*

- An *alert* is a message from the reader, about the health of the reader. If something isn't working, the reader sends an Alert.

- An *observation* is a message from the reader that reports some value or data point on a tag or tags at a particular pre-selected place or time.

- A *command* is a message from the network to the reader. There are *configuration commands,* which cause the reader to set itself up in a particular way. There are *observation commands,* which cause the reader to read or write or modify tag information. There are *trigger commands,* which set the level of event that will cause a particular kind of notification or activity by the reader.

The reader contains the electronic equivalent of a turnstile, preventing collisions when signals from many tags arrive simultaneously. It also has a process to accomplish Singulation, which is the ability to pick out individual tags by serial number.

Environment Issues

RFID works differently with different items in different places. When a system is set up, you have to figure out how your unique situation must be handled.

Item environmentals

What is the item we are trying to track? Is it bottles, metal cans, boxes? Cows? How far away will it be from the Reader? Will it be extremely hot or cold? Are we tracking mixed pallets? What in the environment might damage our tags? What kind of equipment is handling the items, and will it interfere? What kind of changes will we want to make to passing items in response to our reads? They're all common-sense things, but there are a lot of them.

System environmentals

What other sources of noise or interference exist in the place where we want to read tags? How fast are tagged items going? Are there other products right next to them that we don't want to read? Are there big changes in the environment while we are reading? Are there a lot of electric motors (a huge source of radio interference) near where we want to read? Is there a "null spot" where tags cannot pick up enough energy from the reader to operate the chip? You can respond to environment problems with different tags, different tag placement, different antennas, and so on.

Predictions

You can put a tag or a reader or both into a mobile phone. That means individual consumers can use RFID. They can read a tag on a package, poster or shelf and respond to the information. Phones which have a tag or reader or both are already available in Japan and Korea, and will soon be available in the United States.

Several chipmakers and technology providers have produced a "reader on a chip." This should dramatically drive down the price of readers, and put them in a lot more places. It is expected to be the tipping point for many RFID applications.

Middleware

RFID systems collect enormous amounts of data. In order to know where certain items are, in real time, at a particular moment, we track where all the tagged items are, all the time that they are in range of a reader. The current Wal-Mart application, when rolled out across all the stores and all the products it proposes to track, will generate, every three days, more information than is contained in the Library of Congress.[1] It is a thousand times as much data as a bar code system would produce. It would be expensive to carry all that data from the edge to the center, and the ERP at the center could not process it all.

Middleware is the solution to this problem. Middleware is software that lives at the edge of a network. Groups of readers send their observations and events to the middleware application. Middleware filters the data to collect only those observations which have previously been defined as events. It prepares event data so that it can be used by applications upstream from the edge. It routes different portions of the data to different applications. It may also permit applications to redefine events, or request chunks of the data that were not previously being passed along.

Middleware can generate alerts when expected products don't arrive. It can provide alerts when perishable products arrive and must be tended to. Middleware can unify data streams from different kinds of readers. If a reader goes down, the middleware can turn on a back-up. Where reading is imperfect, middleware can do "smoothing" to eliminate the "flickers" where items seem to appear and disappear. Where overlapping antennas generate duplicate reads, middleware can adjust for that. And middleware can organize historical data, not forwarded to the center, so that it can be queried later, if necessary. Finally, if needs change, it's cheaper and easier to modify the middleware than to change the big, central ERP. Perhaps 1 percent of the tag data will be forwarded to the center by middleware. Some middleware functions will migrate to readers.

Business Rules and Event-Driven Responses

Events defined by the middleware and detected by the readers can trigger an instant, automatic response, if Business Rules have been written in advance to tie the response to the event.

Making Business Rules for the customer experience is not easy. But it is being done, and companies get better with practice. Some of the big opportunities may be in areas that marketers don't think about, because without real-time awareness, they would be impossible.

Consider the problem of putting a product "on sale." It's the single most common form of promotion. There are all kinds of rules-of-thumb for how to do it. And all of us do it badly lots of the time. How much should you reduce the price of a product, in order to sell so many more units that you make up for the reduction in margin? Will your price cut be big enough to bring in new users who will generate more full-price sales down the road? Do you need to cut the price a lot more in July than you would a month later at back-to-school? Can you time the sale so that you get new customers, but your regular full-price customers are not just waiting for the sale to happen? Are you cutting the price much more than is necessary to get the extra sales and new users?

Marketers have treated the product-on-sale as a giant, irrevocable experiment. We try to learn from each promotion and apply the experience to the next one, but it is a slow and fallible way to learn, yielding as much folklore as fact.

In the real-time world of RFID, you could automatically watch a sale in each store, moment by moment. If our 30 percent reduction is blowing out the inventory on day one at store ten, maybe we cut back to 20 percent off and see what happens. Not enough action? Try 25 percent. This would be too expensive if people did the watching, but with Business Rules and event-driven responses, not so much.

Business Rules could involve the relationship with key customers. The 19TH century shopkeeper knew when an important customer walked in the door. The twenty-first century bank manager hasn't a clue. An alert to the front-line service provider is an easy application from the design side.

Tagging people creates a lot more issues than tagging products. Nobody wants to walk around with a bullseye on his back. But we've learned from the long struggle with loyalty programs that some of the customers some of the time will feel good about carrying and showing a VIP card. Maybe there's a way to make them feel good about a VIP card that doesn't have to be shown.

Printers, applicators, labels, and packages

The most common way to put a tag on a consumer product is to encode a tag, print a label that carries the tag, and stick the label on the product somewhere toward the end of the production line. There are some issues in doing this right. The software needs to get the right information on the tag to match each different product that is going by. It will be different every time. A verifier reads each tag and makes sure it communicates the right information. The label needs to be at the right spot on the package, with the right read range for the environment. There are adhesion problems and problems with labels that are bent around tight corners and problems with the discharge of static electricity during the application process. However, basically this is a solved problem.

It is not yet possible to print an RFID tag containing a computer chip and an antenna onto a package. But you can print the antenna portion. The printer can paste a chip on to the label in connection with the antenna. There are new systems that encode tags after they are on the package, to attain higher manufacturing speeds.

Standards and protocols: The eco-system of RFID

RFID is about sharing data quickly, efficiently, and in large amounts, across different companies and their customers. Electronic sharing requires shared rules: shared systems for formatting data, transmitting data, encoding data, storing data, and so on. A key strength of RFID is that, early in its development, it was given a robust and flexible set of standards and protocols, which drew on learning from the early days of the Internet. Standards govern every step and every aspect of RFID communication: how tags are labeled, how they arrange data, how data is secured, how it is communicated, how tags couple with readers, how readers interact with tags and middleware, and so on. Standards mean that readers and tags from different manufacturers are interchangeable, within a type. This permits price competition and speeds globalization. There is an Object Naming Service (ONS), with standards for names so that products in the same category are named the same way. A centralized online Discovery Service lets the user look up the information communicated by a tag number: who manufactures it, which version it is, and so on.

Many of the basic standards and protocols were originally developed in the Auto-Identification laboratories at MIT. Standards and protocols are now in the hands of EPCglobal, an independent body supported by manufacturers and users of RFID systems. Standards are developed and voted on by EPCglobal members.

EPCglobal is currently using its second generation of tag standards. The standards define minimum performance requirements for tags, readers and systems. They ensure compatibility across vendors.

Smart cards and NFC as a subset of RFID

As mentioned above, tags and readers can be designed to communicate at a very short distance, or at a longer distance.

When they are designed to communicate close up, a common approach is *magnetic induction*. A current sent through a coil in the reader creates a magnetic field around the coil. When a tag comes inside that field, the tag's coil gets voltage from the reader's coil. In addition to powering the tag, this voltage can be bounced back and modulated to transfer data. The field around the reader's coil is called the *near field*.

When the tag and reader are designed to communicate at a longer distance, the process is different. Again, a current that goes through the coil in the reader creates a magnetic field around the coil. But that current also creates waves that radiate into space beyond the coil. In this design, the tag captures those waves. *Wave capture* permits the transfer of power to the tag, and of information back to the reader. The area outside the magnetic field around the reader's coil, within which waves can be captured, is called the *far field*.

Both methods have the same goals and the same basic idea. But historically, they have been used for different applications. Many of the applications of Near Field Communication (NFC) involve payment systems, where powerful security protocols have been developed. Many of the applications of far field communication involve asset tracking or supply chain management, where security requirements are not high. So users of near field communication see their applications as very different from the rest of RFID. They refer to NFC and RFID as two different things.

This is more about marketing than about technology. The two methods are often in competition; lots of jobs that can be done with Near Field can also be done with a Far Field design. But it can be confusing to the manager. See the two systems as variations on a basic method for automatic identification, each of which has its own strengths and weaknesses.

- NFC is the technology usually used for contactless smart cards that can simply tap a reader to communicate.

- NFC is used in ticketing. Tap-and-enter systems for getting on the subway are a whole lot faster than systems which read a magnetic stripe on your ticket.

- NFC is used to make a quick connection between devices. Like a mouse moving a cursor, it can select an appliance just by getting close to it. It creates a quick, secure, user-friendly handshake, in PCs, cameras, set-top boxes, POS equipment, and vending machines. It will probably be the basic method for moving photos from camera to computer or from camera to printer: you touch the tools to each other to establish a wireless connection.

Contactless smart cards

A contactless smart card is a tag that uses NFC.

Many smart cards today have a microprocessor as well as storage. Some have more than one microprocessor. Smart cards can store (today) up to 512 *kilobytes* of data, which was once the capacity of a whole hard drive. The magnetic stripe cards they replace can store only a few bytes.

A smart card with a processor can manage files, process data, execute algorithms and direct interactive processes. Again, read range is short, to protect privacy. Its data can be password protected. And memory can be partitioned for different applications. That also protects privacy. Data can be protected from access and manipulation. It makes information-based transactions simple, safe, and inexpensive.

Smart cards used in payment systems address the basic design flaw in conventional credit cards, which is that the magnetic stripe doesn't have enough storage capacity for encryption. Thus the data used to authorize payments of thousands of dollars by an individual is accessible to anyone who holds the card even for a few seconds. How can this be a good idea? The smart card has sufficient storage and processing to empower encryption. Visa has estimated that the replacement of conventional cards with smart cards reduces fraud by 70 to 90 percent.[2] In addition to encryption, smart cards support other security features, such as hashes, digital signatures, dynamic data generation, mutual authentication with the reader, built-in capabilities that detect and react to tampering, PIN support, and support for biometrics for human identification.

Mesh Networks

Tags with sensors can be arranged in a mesh network, composed of individual nodes which can communicate with each other. They can be spread over a large area to detect changes in light or temperature or gunshots or radiation or something else, anywhere within the boundaries. They don't have to be wired together—just scattered. It's a cheap and durable way to monitor a big space. In a mesh network, signals can bounce from node to node, so they don't have to travel far. Signals can route around destroyed nodes or dead batteries, to keep the network alive. Power can come from batteries or solar cells. Systems for monitoring wildlife habitat use mesh networks. You can track firemen in a burning building with a mesh network.

Zigbee

Zigbee is another protocol for low data-rate, low power radio signals used in a Mesh Network. It is much like active-tag RFID and fits some of the same applications.

Bluetooth

Bluetooth is a protocol for linking nearby appliances without wires. Bluetooth piconets have one master and seven slaves, maximum. If you have more than eight devices, you have to park some of them to let the others into the network. If many masters are close together, there are collision problems. Where networks are larger, Bluetooth must be replaced by RFID or Zigbee.

RFID as a service

AT&T in the U.S. and Telekom Austria in Europe have launched RFID as a service. It provides tags, readers, software, installation, and a start at integration with the business processes of the user. The early service offerings focus on supply chain management and asset tracking. RFID as a pre-engineered service will open the door to smaller firms, and to those who are concerned about the frontier nature of RFID technology.

Books about RFID Technology

Following is a short list of books that explain RFID technology. Any of them will give you a much more detailed understanding of how its components and processes work.

Finkenzeller, Klaus. *RFID Handbook: Fundamentals and Applications on Contactless Smart Cards and Identification*. Second Edition. Chichester: John Wiley & Sons Ltd., 2003.

Glover, Bill and Himanshu Bhatt. *RFID Essentials*. Sebastopol, CA: O'Reilly Media Inc., 2006.

Shepard, Steven. *RFID: Radio Frequency Identification*. New York: McGraw-Hill, 2005.

Want, Roy. *RFID Explained: A Primer on Radio Frequency Identification Technologies*. San Francisco: Morgan & Claypool, 2006.

Notes

Chapter 1

1. Golan, Elise, Barry Krissof, Fred Kuchler, Linda Calvin, Kenneth Nelson, and Gregory Price. "Traceablity in the US Food Supply: Economic Theory and Industry Studies," *Agricultural Economic Report* Number 830. United States Department of Agriculture. HD9005. Published in Rasco, Barbara A. and Glen E. Bledsoe. *Bioterrorism and Food Safety.* New York: CRC Press, 2005. Page 361.

2. Gregory, Jonathan. "RFID and SAP: A Strategic Vision." White paper, Computer Sciences Association. May 2006. Passim. See also Heinrich, Claus. *RFID and Beyond.* Indianapolis: Wiley Publishing, Inc., 2005. Passim.

3. Hughes, Sandy. "P&G: RFID and Privacy in the Supply Chain." In Simpson, Garfinkel and Beth Rosenberg, editors. *RFID: Applications, Security and Privacy,* pp. 404-405. New York: Addison-Wesley, 2006.

4. Bacheldor, Beth. "China Post Deploys EPC RFID System to Track Mailbags," *RFID Journal,* 13 July 2006. www.rfidjournal.com/article/articleprint/2487/-1/1/

5. Sullivan, Laurie. "RFID Rides High on Jet Engines." *Information Week,* 20 April 2005. www.informationweek.com/news/mobility/RFID/showArticle.jhtml?articleID=160910603

6. Gamble, Paul, Merlin Stone and Neil Woodcock. *Up Close and Personal: Customer Relationship Marketing at Work.* London: Kogan Page, 1999. Page 226.

7. Kasanoff, Bruce. *Making It Personal: How to Profit From Personalization Without Invading Privacy.* Cambridge, MA: Perseus Publishing, 2001. Page 104.

8. Peppers, Don and Martha Rogers. *Managing Customer Relationships: A Strategic Framework.* Hoboken, NJ: John Wiley & Sons, 2004. Page 3.

9. Ibid., page 4.

10. Clegg, Brian. *Capturing Customers' Hearts.* Upper Saddle River, NJ: Prentice Hall, 2000. Page 3.

11. Sarma, Sanjay. "A History of the EPC" in Garfinkel, Simpson and Beth Rosenberg, eds. *RFID: Applications, Security and Privacy.* Upper Saddle River, NJ: Addison-Wesley, 2006. Page 39.

12. Ibid., page 41.

13. Joshi, Y. "Information Visibility and its Effect on Supply Chain Dynamics" quoted in Sarma, Sanjay. "A History of the EPC." in Garfinkel, Simpson and Beth Rosenberg, eds. Op. cit. Page 41.

14. Collins, Jonathan. "RFID's Impact at Wal-Mart Greater than Expected." *RFID Journal,* www.rfidjournal.com/article/view/2314/1/1

15. Pandell, Alexander J. "The Alchemist's Wine Perspective." www.wineperspective.com

16. Weiser, Marc. "The Computer for the 21st Century." *Scientific American,* September 1991, v. 265, n. 3, pp. 94-104. www.ubiq.com/hypertext/weiser/SciAmDraft3.html

Chapter 2

1. Personal Interview: George Dittman, Senior Director for Customer Insights. Harrah's Entertainment, Inc.

2. Buttle, Francis. *Customer Relationship Management: Concepts and Tools.* Burlington, MA: Elsevier Butterworth Heinemann, 2004. Page 236.

3. Ibid., page 127.

4. Ibid., page 119.

5. Among many, see Cram, Tony. *Customers that Count.* New York: Prentice-Hall, 2001. Passim.

6. Ibid., page 91.

7. Newell, Frederick. *Why CRM Doesn't Work: How to Win by Letting Customers Manage the Relationship.* Princeton, NJ: Bloomberg Press, 2003. Page 159.

8. Barnes, James G. *Secrets of Customer Relationship Marketing.* New York: McGraw-Hill, 2001. Page 157.

9. Swedberg, Claire. "RFID System Helps Retailers Send Tailored Messages to Customers." *RFID Journal,* www.rfidjournal.com/article/articleview/3024/

Chapter 3

1. First formulated in Greenberg, Paul. CRM *at the Speed of Light: Capturing and Keeping Customers in Internet Real Time.* New York: McGraw-Hill, 2002. Page 16.

2. Cram, Tony. *Customers that Count: How to Build Living Relationships with Your Most Valuable Customers.* London: Pearson/Prentice Hall, 2001. Page 3.

3. Reichheld, Frederick F. *The Loyalty Effect.* Boston: Harvard Business School Press, 1996. Page 48.

4. Swift, Ronald S. *Accelerating Customer Relationships: Using CRM and Relationship Technologies.* Upper Saddle River, NJ: Prentice Hall, 2001. Page 79.

5. Schloter, Philip and Hamid Aghajan. "Wireless RFID Networks for Real-Time Customer Relationship Management." Paper presented at the First International Workshop on RFID and Ubiquitous Sensor Networks. Nagasaki, Japan. December 2005. www.informatik.uni-trier.de/~ley/db/conf/euc/eucw2005.html

6. Buttle, Francis. *Customer Relationship Management: Concepts and Tools.* Burlington, MA: Elsevier Butterworth Heinemann, 2004. Page 308.

7. Swedberg, Claire. "RFID System Helps Retailers Send Tailored Messages to Customers." *RFID Journal,* 6 February 2007. www.rfidjournal.com/article/3024/-1/1/

8. Hotchkiss, D'Anne. "New TeraData Retail Advanced Business Analytics & RFID Lab Delivers Insight from Enterprise." www.teradata.com/t/page/150760/index.html

9. Buttle, Francis. Op. cit., pp. 124–127.

10. Griffin, Jill and Michael W. Lowenstein. *Customer Winback: How to Recapture Lost Customers and Keep Them Loyal.* San Francisco: Jossey Bass Publishing, 2001. Pages 23, 110.

11. Griffin, Jill and Michael W. Lowenstein. Op. cit. Pages 53-88.

12. Peppers, Don and Martha Rogers. *Managing Customer Relationships.* Hoboken, NJ: John Wiley & Sons, 2004. Pages 137–148.

13. Peppers, Don and Martha Rogers. Op. cit., page 20.

14. Bacheldor, Beth. "Health Facility Uses RTLS to Provide Concierge Care." *RFID Journal,* 9 October 2007.

15. Reichheld, Frederick F. (with Thomas Teal.) *The Loyalty Effect: The Hidden Force Behind Growth, Profits and Lasting Value.* Boston: Harvard Business School Press, 1996. Page 36.

16. Cram, Tony. Op. cit. Page 187.

17. _____. "Radio Frequency Identification: Moving Beyond the Hype to Maximum Value". White paper, UNISYS. 2004. Downloaded on 28 August 2008 at: www.rfidc.com/pdfs_downloads/Unisys%20RFID%20White%20Paper.pdf

18. Cram, Tony. Op. cit. Page 16.

Chapter 4

1. Newell, Frederick. *Why CRM Doesn't Work: How to Win by Letting Customers Manage the Relationship.* Princeton, NJ: Bloomberg Press, 2003. Page 112.

2. Thompson, Bob. "Customer Experience Management: The Value of 'Moments of Truth.'" White paper, Crmguru.Com and Rightnow Technologies. 2006. Downloaded on 28 August 2008 at: crm.rightnow.com/cgi-bin/rightnow.cfg/php/enduser/doc_serve.php?2=WPCRM-FORM-CRMX-change-TextAd-070903-CEMI

3. Thompson, Bob. "Customer Experience Management: Accelerating Business Performance," part 2 of 2. White paper, Customerthink Corporation and Rightnow Technologies. Downloaded on 28 August 2008 at: www.retaintogain.com/pdf/customer_exp2.pdf

4. Ibid.

5. _____. "Radio Frequency Identification: Moving Beyond the Hype to Maximum Value." White paper, UNISYS. 2004. Downloaded on 28 August 2008 at: www.rfidc.com/pdfs_downloads/Unisys%20RFID%20White%20Paper.pdf

6. Schloter, Philip and Hamid Aghajan. "Wireless RFID Networks for Real-Time Customer Relationship Management." Paper presented at the First International Workshop on RFID and Ubiquitous Sensor Networks, Nagasaki, Japan. December 2005. Downloaded on 28 August 20008 at: www.informatik.uni-trier.de/~ley/db/conf/euc/eucw2005.html

7. Ibid.

8. Newell, Frederick and Katherine Newell Lemon. *Wireless Rules: New Marketing Strategies for Customer Relationship Management Anytime, Anywhere.* New York: McGraw Hill, 2001. Page 255.

9. _____. "Wirelessly Extending RFID Tracking Systems." CAL AMP. www.calamp.com

10. Brown, Stanley A. *Customer Relationship Management: A Strategic Imperative in the World of E-Business.* Toronto: John Wiley & Sons Canada, 2000. Pages 41–48.

11. _____. "Real Time Promotion Execution." White paper, OATsystems. 2005.

12. Gregory, Jonathan. "RFID ands SAP: A Strategic Vision." White paper, Computer Sciences Corporation. May 2006. Downloaded on August 28, 2008 at: www.csc.com/aboutus/leadingedgeforum/knowledgelibrary/uploads/1128_1.pdf

13. Greenfield, Adam. *Everyware: The Dawning Age of Ubiquitous Computing.* Berkeley, CA: New Riders (Pearson Education), 2006. Pages 227–260.

Chapter 5

1. Clarke, R. A. "Human Identification in Information Systems: Management Challenges and Public Policy Issues." *Information Technology and People.* December 1994, v. 7, issue 4, pp. 6–37. Downloaded August 28, 2008 at: www.anu.edu.au/people/Roger.Clarke/DV/HumanID.html

2. Solove, Daniel J. *The Digital Person: Technology and Privacy in the Information Age.* New York: New York University Press, 2004. Page 33.

3. Hanson, David. "Fake ID Cards." *Alcohol: Problems and Solutions.* Downloaded on August 28, 2008 at: www2.potsdam.edu/hansondj/YouthIssues/1048597141.html

4. Harper, Jim. *Identity Crisis: How Identification Is Overused and Misunderstood.* Washington, D.C.: Cato Institute, 2006. Page 122.

5. Dash, Eric. "Stolen Lives." *New York Times,* 12 December 2006. Downloaded on 28 August 2008 at: www.nytimes.com/2006/12/12/business/12credit.html

6. Harper, op. cit. Page 189.

7. Harper, op. cit. Passim

8. _____. "Logical Access Security: The Role of Smart Cards in Strong Authentication." White paper, Smart Card Alliance. Downloaded on 28 August 2008 at: www.smartcardalliance.org/pages/publications-logical-access-report

9. www.flyclear.com/

10. Gelfand, Alexander. "Startup Plans to Solve Online Identity Theft, But Does Anyone Care?" *Wired,* 8 February 2008. Downloaded on August 28, 2008 at: www.wired.com/politics/security/news/2008/02/credentica

11. Collins, Jonathan. "E-passport Tag comes with Switch." *RFID Journal,* 23 May 2006.

12. "Logical Access Security: The Role of Smart Cards in Strong Authentication." Op. cit.

13. Espiner, Tom. "Cracking Open the Cybercrime Economy." ZDNet (UK) Downloaded on 28 August 2008 at: http://news.zdnet.com/2100-1009_22-6222896.html

14. _____. "RFID and Consumers: Understanding Their Mindset: A U.S. Study Examining Consumer Awareness and Perceptions of Radio Frequency Identification Technology." Page 7. Downloaded on 28 August 2008 at: www.us.capgemini.com/DownloadLibrary/requestfile.asp?ID=400

15. Gunther, Oliver and Sara Spiekerman. "RFID and the Perceptions of Control: The Consumer's View." Communications of the ACM 2005, v. 48, n. 9, pp. 73–76.

16. Taylor, Humphrey. "Most People Are 'Privacy Pragmatists' Who, While Concerned about Privacy, Will Sometimes Trade It Off for Other Benefits." The Harris Poll #17. March 19, 2003. Downloaded on 28 August 2008, at: www.harrisinteractive.com/harris_poll/index.asp?PID=365 (Note: This is one in a series of eight nearly identical studies; all had the same outcome.)

17. Mosquera, Mary. "Ranks of Privacy Pragmatists are Growing." TechWeb 7 December 2000. Downloaded on August 28, 2008 at: www.techweb.com/wire/story/TWB20001207S0002

18. Collins, Jonathan. "RFIDsec Unveils Privacy-Protected Tags." RFID Journal, 21 June 2006.

19. Ohkubo, Miyako and Koutarou Suzuki and Shingo Kinoshita. "RFID Privacy Issues and Technical Challenges." Communications of the ACM, September 2005, v. 48, n. 9, pp. 66–71.

20. Swedberg, Claire. "Marnlen Makes Privacy-Friendly Tags for Retail Items." RFID Journal, 8 November 2006.

21. O'Connor, Mary C. "New Tag Aims to Protect Consumer Privacy." RFID Journal, 2 April 2007.

22. _____. "Understanding Gen2" What it is, How you will benefit and criteria for vendor assessment." White paper, Symbol Technologies. January 2006. Downloaded on 28 August 2008 at: www.symbol.com/1521/18

23. Collins, Jonathan. "E-passport Tag comes with a Switch." RFID Journal, 23 May 2006.

24. Zou, Cliff C. "PCB: Physically Changeable Bit for Preserving Privacy in Low-End RFID Tags." White paper, University Of Central Florida. Undated. Downloaded on 28 August 2008 at: www.cs.ucf.edu/~czou/research/PCB.pdf

25. Pearson, Joseph. "An RFID Tag Data Security Infrastructure Approach for Items." RFID Journal, 18 December 2006.

26. Nohara, Yasunobu and Suzo Inoue, Kensuki Baba, Hiroto Yasuura. "Quantitative Evaluation of Unlinkable ID Matching Schemes." Proceedings of the 2005 ACM Workshop On Privacy and the Electronic Society. Downloaded on 28 August 2008 at: http://portal.acm.org/citation.cfm?id=1102212

27. Peppers, Don and Martha Rogers. "Anonymization Offers New Approach to Data Sharing." 1:1 Privacy, 14 September 2006.

28. "Logical Access Security: The Role of Smart Cards in Strong Authentication." Op. cit.

29. Willoughby, Mar. "RFID Security Worries Need a Reality Check." Computer World, 1 May 2006.

30. Garfinkel, Simpson. "An RFID Bill of Rights." Technology Review, October 2002, v. 105, issue 8.

31. _____. "The Use of RFID for Human Identity Verification." RFID Subcommittee Report, Department of Homeland Security. Draft: 12/6/2006. Downloaded on 28 August 2008 at: www.dhs.gov/xlibrary/assets/privacy/privacy_advcom_12-2006_rpt_RFID.pdf

32. _____. "RFID Overview." White paper, Caslon Analytics. Undated. Downloaded on 28 August 2008 at: www.caslon.com.au/rfidprofile.htm.

33. Stone, Brad. "Is that a Radio in Your Cereal?" Newsweek, 29 September 2003, v. 142, issue 13.

34. _____. "RFID Privacy Page: Radio Frequency Identification Systems." Electronic Privacy Information Center. Downloaded 28 August 2008 at: http://epic.org/privacy/rfid/

35. Ibid.

36. _____. "RFID and Consumers: Understanding Their Mindset: A U.S. Study Examining Consumer Wareness and Perceptions of Radio Frequency Identification Technology." Page 7. Downloaded 28 August, 2008 at: www.us.capgemini.com/DownloadLibrary/requestfile.asp?ID=400

37. _____. "Radio Frequency Patient Monitoring: A Cost-Effective patient/Staff model for the Emergency Department." White paper, RFT Technologies. Undated. Downloaded on 28 August 2008 at: www.rft.com/templatefiles/includes/common/displayFile.ashx?fileId=197

38. Swedberg, Claire. "AudioTel Uses RFID to Protect Financial Data." *RFID Journal*, 26 July 2006.

39. "RFID Overview." Op. Cit. See also _____. "RFID Privacy Page: Radio Frequency Identification Systems" Op. cit.

40. _____. "RFID and Consumers: Understanding Their Mindset. Op. cit.

Chapter 6

1. ____. "Change is in the Air." *The Economist*, 12 March 2005, v. 374, issue 8417.

2. Collins, Jonathan. "Air France-KLM Embarks on RFID Luggage-tag Trial." *RFID Journal*, 18 August 2006.

3. McCartny, Scott. "A New Way to Prevent Lost Luggage." *Wall Street Journal Online*, 27 February 2007. Downloaded 28 August 2008 at: www.volweb.cz/horvitz/os-info/news-feb07-028.html

4. Ibid.

5. _____. "The Case for Smart Luggage" *Science Spectra* 2000, Issue 21.

6. Wyld, David C. "The Wide World of Sports Evolves via RFID." *Contactless News*, 20 June 2006. Downloaded 25 August 2008 at: www.contactlessnews.com/2006/06/20/the-wide-world-of-sports-evolves-via-rfid/

7. Transit Cooperative Research Program Report 108. "Car Sharing: Where and How it Succeeds" White Paper: Federal Transit Administration. Downloaded 25 August 2008 at www.nelsonnygaard.com/articles/article_carsharing.htm

8. Heinrich, Claus. *RFID and Beyond*. Indianapolis: Wiley Publishing, Inc., 2005. Page 146.

9. _____. "Near Field Communication in the Real World." White paper, Innovision. Undated. Downloaded 28 August 2008 at www.nfc-forum.org/resources/white_papers/Innovision_whitePaper1.pdf

10. Laursen, Wendy. "Managing the MegaFlock." *IEEE Review*, February 2006., v. 52, issue 2, pp. 38-42.

11. Heinrich, op. cit., chapters 1 and 2.

12. Wessel, Rhea. "German Baker's RFID Application is Recipe for Success." *RFID Journal*, 28 Aug 2006.

13. Finkenzeller, Klaus. *RFID Handbook: Fundamentals and Applications in Contactless Smart Cards and Identification*, second edition. Chichester: John Wiley & Sons Ltd., 2003. Page 363.

14. _____. "CHEP Begins RFID Deployment for Its Container Fleet." RFIDUpdate.com 15 May 2007. Downloaded on 28 August 2008 at: www.rfidupdate.com/articles/index.php?id=1233

15. Finkenzeller, Klaus. Op. cit., pp. 376–377.

16. _____. "Tulsa To Deploy RFID Automated Bike Rental Rack." RFIDupdate.com. Downloaded on 28 August 2008 at: www.rfidupdate.com/articles/index.php?id=1373

17. Greengard, Samuel. "Mississippi Blood Services Banks on RFID." RFID Journal, 7 August 2006.

18. ____. "RFID Overview." White paper, Caslon Analytics, undated. Downloaded 28 August 2008 at: www.caslon.com.au/rfidprofile.htm

19. Ayre, Lori Bowen. "Wireless Tracking in the Library: Benefits, Threats and Responsibilities." In Garfinkel, Simpson and Beth Rosenberg, eds. *RFID: Applications, Security and Privacy*. Upper Saddle River, NJ: Addison-Wesley, 2006. Page 235.

20. _____. "Car Dealership Finds RFID the Key to Increased Sales." *RFID Journal*, 27 July 2006.

21. ____. "Sealed with a Tag." *RFID Journal*, 18 February 2008.

22. Wasserman, Elizabeth. "Construction's Building Blocks: RFID." *RFID Journal*, 16 February 2007.

23. Wasserman, Elizabeth. "Cashing in on RFID's Benefits." *RFID Journal*, 16 February 2007.

24. Neutkens, Debra. "RFID In, Variation Out." *National Hog Farmer*, 15 August 2005.

25. O'Connor, Mary C. "RFID Cures Concrete." *RFID Journal*, 30 October 2006.

26. Read, Reik, Rob Timme, and Samantha Delay. "Supply Chain Technology: RFID Monthly." White paper, Robert W. Baird & Company, June 2007.

27. Nobel, Carmen. "Symbol Considers RFID Options." *E-Week*, 13 February 2006.

28. Swedberg, Claire. "RFID Markers Track Buried Cables at Atlanta Airport." *RFID Journal*, 12 September 2006.

29. _____. "Beyond the tag: Finding RFID Value in Manufacturing and Distribution Applications." White paper, Intermec. 2004. Downloaded 28 August 2008 at: epsfiles.intermec.com/eps_files/eps_wp/beyondthetag_wp_web.pdf

30. Swedberg, Claire. "DHL Demos RFID-Enabled Delivery Van." *RFID Journal*, 6 March 2007.

31. Overby, Christine S. and Ellen Daley. "Topic Overview: The Extended Internet." Forrester Research. 4 April 2006.

32. Borriello, Gaetano , Waylan Brunette, Matthel Hall, Carl Hartung, and Camerson Tangney. "Reminding About Tagged Objects Using Passive RFID." Downloaded 28 August 2008 at: www.uwnews.org/relatedcontent/2004/October/rc_parentID5748_thisID5749.pdf

33. Bacheldor, Beth. "Telepathx Develops Accident-Detecting System Linked to Auto Airbags." *RFID Journal*, 1 August 2007.

34. O'Connor, Mary C. "RFID Tidies Up the Distribution of Scrubs." *RFID Journal*, 6 February 2007.

35. Swedberg, Claire. "Wisconsin Clinic Opts for RFID Solution." *RFID Journal*, 28 December 2006.

36. O'Connor, Mary C. "UWB Vendor Announces Pure RFID Platform." *RFID Journal*, 20 December 2006.

37. Swedberg, Claire. "ExploTrack Launches e-Pedigree Platform for Explosives" *RFID Journal*, 6 June 2006.

38. Finkenzeller, Klaus. Op. cit., pp. 371–376.

39. Ferguson, Glover T. "Have Your Objects Call My Objects." *Harvard Business Review*, June 2002, v. 80, i. 6.

40. _____. "An Automated RFID Solution for Physical IT Asset management and Protection." White paper, AXCESS International 2006. Downloaded 28 August 2008 at: parts.ihs.com/news/dupes/axcess-white-paper.htm

41. Swedberg, Claire. "AudioTel Uses RFID to Protect Financial Data." *RFID Journal*, 26 July 2006.

42. Wasserman, Elizabeth. "Cashing in on RFID's Benefits." *RFID Journal*, 16 February 2007.

43. O'Connor, Mary C. "Imation Announces RFID Solution for Tracking Data Tapes." *RFID Journal*, 8 March 2007.

44. O'Connor, Mary C. "Packaging Maker Offering Tamper-Evident RFID Film." *RFID Journal*, 10 January 2007.

45. Hanebeck, Christian, "Building the Intelligent Network of Collected Items." White paper, Globe Ranger, 2006.

46. Swedberg, Claire. "Port of Oakland Sees Signs of Security in RFID." *RFID Journal*, 15 March 2007.

47. Swedberg, Claire. "Dutch Horticultural Company Sends Flowers via RFID." *RFID Journal*, 3 November 2006.

48. O'Connor, Mary C. "IPICO Enters Race for RFID Sports Timing." *RFID Journal*, 14 March 2007.

49. O'Connor, Mary C. "IT Cabling Company Offering RFID Value-Add." *RFID Journal*, 15 February 2007.

50. _____. "Apple Patent Uses RFID for Home Networking." RFIDupdate.com 9 March 2007. Downloaded 28 August 2008 at: www.rfidupdate.com/articles/index.php?id=1314

51. Want, Roy. RFID *Explained: A Primer On Radio Frequency Identification Technologies*. Morgan & Claypool, 2006. Pages 42–43.

52. Gambon, Jill. "Ultra-Wideband RFID Tracks Nuclear Power Plant Workers." *RFID Journal*, 8 March 2007.

53. Haag, Werner R. and Durstenfield, Bob, and Fuhr, Peter, et. al. "Cargo Container Security." August 2003. Downloaded on 28 August 2008 at: www.ohsonline.com/articles/44589/

54. Bacheldor, Beth. "Fighting Fires with RFID and Wireless Sensors." *RFID Journal*, 7 November 2006.

55. Swedberg, Claire. "New RFID System Takes Security to Heart." *RFID Journal*, 23 March 2007.

56. Peeters, John P. "Ask the Experts: John Peeters, President and Founder, Gen Tag." *Contactless News*, 3 April 2005. Downloaded 28 August 2008 at: www.contactlessnews.com/library/2005/04/03/ask-the-experts-john-peeters-president-and-founder-gentag/

57. _____. "Starbucks Keep Fresh with RFID." *RFID Journal*, 13 December 2006.

58. Swedberg, Claire. "RFID Helps Keep Avocados Fresh." *RFID Journal*, 2 April 2007.

59. _____. IBM Sensor and Actuator Solutions. White paper, Downloaded 28 August 2008 at: www-03.ibm.com/solutions/businesssolutions/sensors/doc/content/bin/S_A_overview_FINAL.pdf
60. (auburn.edu/audfs) Cited in Want, Roy. Op. cit., page 45.
61. Wessel, Rhea. "Pharma Label Maker to Test Tags that Record Temp." *RFID Journal*, 25 September 2006.

Chapter 7

1. _____. "Item-level RFID technology redefines retail operations with real-time collaborative capabilities." White Paper, IBM, 2004. Downloaded 28 August, 2008 at: www-03.ibm.com/solutions/businesssolutions/sensors/doc/content/bin/RFID_eBrief_ Final_2a.pdf
2. Ibid.
3. Ferreira, Devon and Girish Ramachandran. "Simplify: Leveraging RFID and Other Pervasive Technologies to Achieve Transaction Efficiency in Customer Retailing." White paper, Information Technology, April 2006. Downloaded August 28, 2008 at: www.infosys.com/RFID/simplify-consumer-retailing.pdf
4. Collins, Jonathan. RFID's Impact at Wal-Mart Greater than Expected." *RFID Journal*, 4 May 2006
5. Wasserman, Elizabeth. "RFID is in Fashion." *RFID Journal*, 19 February 2007.
6. Embry, Wayne. "Are you ready for RFID?" White Paper, SAS, Undated. Downloaded 28 August 2008 at: www.sas.com/ci/whitepapers/102274.pdf
7. Wasserman, Elizabeth. "RFID is in Fashion." *RFID Journal*, 19 February 2007.
8. Salmon, Kurt. "Moving Forward with Item-Level Radio Frequency Identification in Apparel/Footwear." White Paper, Kurt Salmon Associates, 2005. Downloaded 28 August 2008 at: www.apparelandfootwear.org/UserFiles/File/RFID/White_Paper-VICS_AAFA_RFID_v11.pdf
9. Hotopf, Max. "The Day of the RFIDs." *Routes to Market Journal*, 19 October 2003. www.the-rtma.com.
10. Personal interview: Marti Barletta, author of *Marketing to Women*. New York: Kaplan Publishing, 2006.
11. Want, Roy. *RFID Explained: A Primer on Radio Frequency Identification Technologies.* Morgan & Claypool, 2006. Page 51.
12. Fildes, Jonathan. "Chips with everything." *New Scientist*, 19 October 2002, v. 176, issue 2365.
13. Swedberg, Claire. "Retailers Test RFID Smart Tables." *RFID Journal*, 7 March 2007.
14. Campo, K., Gijsbrechts, E. and Nisol, P. "The Impact of Retailer Stockouts on Whether, How Much, and What to Buy." *International Journal of Research in Marketing.* 2003, v. 20, pp. 273–286. See also Gruen, Thomas W., Corsten, Daniel S., Bharadwaj, Sundar. "Retail Out of Stocks: A Worldwide Examination of Extent, Causes and Consumer Responses." *Goizueta Paper Series*, 18 April 2002. Downloaded 28 August 2008 at: g bspapers.library.emory.edu/archive/00000035/
15. Brown, Alan S. "REID (sic)Identifies its Business Case." *Mechanical Engineering*, February 2006, v. 128, issue 2.
16. Embry, Wayne. Op. cit.
17. O'Connor, Mary C. "Media Providers Entertain RFID's Potential." *RFID Journal*, 23 June 2006.
18. Heinrich, Claus. *RFID and Beyond.* Indianapolis: Wiley Publishing, Inc., 2005. Page 68.
19. Ibid., page 68.
20. _____. "Real Time Promotion Execution." White paper, OATsystems, 2005.
21. Ibid
22. Collins, Jonathan. "P&G Finds RFID Sweet Spot," *RFID Journal*, 3 May 2006.
23. "Real Time Promotion Execution," op. cit.
24. Collins, Jonathan, op. cit.
25. "Real Time Promotion Execution," op. cit.
26. Underhill, Paco. *Why We Buy: The Science of Shopping.* New York: Simon & Schuster, 2000. Page 189.
27. Fitzgerald, M. *Computing*, 4 March 2004. Page 24. Quoted in Rasco, Barbara A. and Glen E. Bledsoe, op. cit.

28. Lacy, Sarah. "Inching Toward the RFID Revolution." *Businessweek Online* 8/31/2004. Downloaded 28 August 2008 at: www.businessweek.com/technology/content/aug2004/tc20040831_4930_tc172.htm

29. Ferreira, Devon and Girish Ramachandran, op. cit.

30. Swedberg, Claire. "FamilyMart Demonstrates RFID's Convenience to Customers." *RFID Journal*, 2 August 2007.

31. Wasserman, Elizabeth. "RFID Is in Fashion." *RFID Journal*, 19 February 2007.

32. Stone, Chad, Ruben Garcia, Alfonso Lujan, Ben Zoghi and Rohit Singhal. "RFID Solutions: The Custom Specialty Garment Industry." Downloaded 28 August 2008 at: www.idspackaging.com/Common/Paper/Paper_258/RFID%20Solutions.htm

33. Peeters, John P. "Ask the Experts: John Peeters, President & Founder, Gen Tag." *Contactless News*, 3 April 2005. Downloaded August 28 at: www.contactlessnews.com/library/2005/04/03/ask-the-experts-john-peeters-president-and-founder-gentag/

34. _____. "Produce Grower taps Manhattan Associates." *Contactless News*, 6 June 2005. Downloaded 28 August 2008 at: www.contactlessnews.com/2005/06/06/produce-grower-taps-manhattan-associates

35. Betts, Mitch. "RFID Reality Check." *Computerworld*, 20 December 2004. Downloaded 28 August 2008 at: www.computerworld.com/special_report/000/000/700/special_report_000000720_primary_article.jsp

36. Ferguson, Glover T. "Have Your Objects Call My Objects." *Harvard Business Review*, June 2002, v. 80, issue 6.

37. Wood, Christina. "Implications: Tag It." *PC Magazine*, 2 December 2002, v. 21, issue 3.

38. Want, Roy. Op. cit., page 50.

39. Swedberg, Claire. "Philips, SK Telecom to Test NFC in Seoul." *RFID Journal*, 26 May 2006.

40. O'Connor, Mary C. "HP Spots New Opportunities for Passive RFID." *RFID Journal*, 18 July 2006.

41. Swedberg, Claire. "Philips, SK Telecom to Test NFC in Seoul." Op. cit.

42. Bacheldor, Beth. "BRIDGE Expects to Launch Five European RFID Pilots this Fall." *RFID Journal*, 24 July 2007.

43. Ferguson, Glover T. Op. cit.

44. Newell, Frederick. *Why CRM Doesn't Work: How to Win by Letting Customers Manage the Relationship.* Princeton, NJ: Bloomberg Press, 2003. Page 215.

45. Levinson, Meredith. "The RFID Imperative." *CIO Magazine*, December 2005.

46. Zimmerman, Ann. "Retail Losses Rise as Shoplifters go High-Tech." *Pittsburg Post-Gazette*, October 25, 2006. Downloaded 28 August 2008 at: www.post-gazette.com/pg/06298/732855-28.stm

47. O'Connor, Mary Catherine. "RFID at the Car Wash." *RFID Journal*, 25 April 2007.

48. _____. "NEC announces anew CRM Solution." *RFID Gazette*, January 10, 2006. Downloaded 28 August, 2008 at: www.rfidgazette.org/2006/01/nec_announces_a.html

Chapter 8

1. Gambon, Jill. "Florida to Require RFID Tagging for Exotic Pets." *RFID Journal*, 12 February 2007.

2. Athavaley, Anjali. "What Price Green: For Many Americans, Pretty Much any Price Is too High." *The Journal Report: WSJ*, 29 October 2007, page R6.

3. Ibid.

4. Greenfield, Adam. *Everyware: The Dawning Age of Ubiquitous Computing.* Berkeley, CA: New Riders (Pearson Education), 2006. Page 218.

5. Container Recycling Institute. www.container-recycling.org/zbcwaste/recycling.htm

6. Bacheldor, Beth. "RFID-enabled Vending Machine Dispenses Bottled Water." *RFID Journal*, 18 July 2007.

7. Earth911. http://earth911.org/blog/2006/05/24/earth-911-joins-with-national-paint-and-coatings-association-npca-product-stewardship-institute-psi-to-launch-paint-wise-portal/

8. Colorado Recycles. www.colorado-recycles.org/referencelibrary/recyclingtips/paintrevised.
htm#potentialhazards

9. Sarr, Steven and Valerie Thomas. "Toward Trash that Thinks: Product Tags for Environmental Management." *Journal of Industrial Ecology*, 2003, v. 6, n. 2, pp 133–146.

10. Lych, Jim. "Islands in the Waste Stream." Downloaded 28 August 2008 at: www.reciclatic.net/.../
Islands_in_the_Wastestream_Noncommercial_Computer_Reuse_in_the_US_Jim_Lych.ppt

11. www.idealbite.com/tiplibrary/archives/tread_lightly_reducing_tire_waste

12. Rasco, Barbara A. and Bledsoe, Gleyn E. *Bioterrorism and Food Safety*. New York: CRC Press, 2005.

13. Swedberg, Claire. "RFID Helps Reward Consumers for Recycling." *RFID Journal*, 22 February 2008.

14. Themozhi, Jothi S. and Gail Johnson. "The Role of Technology in Creating a Sustainable Community:
Application of Radio Frequency Identification Technology in Curbside Recycling" in Daniels, Mark R.
(ed.) *Creating Sustainable Community Programs*. Westport, CN: Praeger, 2001.

15. Personal Interview: Anthony Romano. Marketing Director, Sonrai. 24 June 2008.

16. Lund, Herbert F. *McGraw-Hill Recycling Handbook*, Second edition. New York: McGraw-Hill, 2001. Page
129.

17. Want, Roy. *RFID Explained: A Primer On Radio Frequency Identification Technologies*. Morgan & Claypool,
2006. Page 64.

18. _____. "NXP and Kestrel Partner on RFID AntiTheft for DVDs" RFIDupdate.com Downloaded 28 August
2008 at: www.rfidupdate.com/articles/index.php?id=1357

19. Sarr and Thomas, op. cit.

20. Pimentel, David and Gardner, Jennifer, Bonnifield, Adam, et. al. "Energy Efficiency and Conservation for
Individual Americans." Springer Science and Business Media. 2007.

21. Greenland, Jr. John. "Go Green." *RFID Journal*, 20 August 2007.

22. Schoenherr, J.I., M. Reichman and T.A. Baloun. "RFID-Based Waste Recycling" White paper, White Paper
Library: *RFID Journal*, 2008.

23. O'Connor, Mary C. "Greek RFID Pilot Collects Garbage." *RFID Journal*, 15 January 2007.

24. Edwards, John. "Most Innovative Use of RFID: Getting the Bugs Out." *RFID Journal*, 1 June 2007. Downloaded 28 August 2008 at:
www.rfidjournal.com/article/articleview/211?Redirect=/article/articleview/3376

25. Songini, Mark. "Colorado Hopes RFID Can Protect Elk Herds." *Computerworld*, 16 January 2006. Downloaded 28 August 2008 at:
www.computerworld.com/mobiletopics/mobile/story/0,10801,107777,00.html

26. Rajewski, Genevieve. "Not Just for Retailers, RFID Helps Track Rainforest Wildlife." *Wired*, 28 June 2007.
Downloaded 28 August 2008 at: www.wired.com/gadgets/miscellaneous/news/2007/06/rfid_pigs

27. Troyk, Philip. "Injectible Electronic Identification Monitoring and Stimulation Systems" *Annual Review of
Biomedical Engineering*, 1999, v. 1, pp 177–209.

28. Culler, David E. "Smart Sensors to Network the World." *Scientific American*, June 2004.
www.intel.com/research/exploratory/smartnetworks.htm

29. McDonough, William and Michael Braungart. *Cradle to Cradle: Rethinking the Way We Make Things*. New
York: North Point Press, 2002. Passim.

30. Shanmugam, Rajavisankar. "RFID—Tackling Product Lifecycle Management Issues." White paper, Tata
Consultancy, September 2004. White Paper Library: *RFID Journal*,

31. MacFarlane, Duncan. "The Living Product." *RFID Journal*, 5 February 2008.

32. Personal Interview: Willie Cade, President. Chicago Computers for Schools.

33. Darby, Sarah. "The Effectiveness of Feedback on Energy Consumption." Downloaded on 28 August
2008 at: www.defra.gov.uk/environment/climatechange/uk/energy/research/pdf/energy
consump-feedback.pdf

Chapter 9

1. Talbot, David. "Where's the Beef From?" *Technology Review*, June 2004, pp. 48–56.

2. Wyld, David C. "Mad Cow Policy: The National Animal Identification System." *Contactless News*, July 21, 2005.

3. Talbot, David, op. cit.

4. Rasco, Barbara A. and Bledsoe, Gleyn E. *Bioterrorism and Food Safety.* New York: CRC Press, 2005.

5. Talbot, David, op. cit.

6. Erdos, Marlena. "RFID and Authenticity of Goods." In Garfinkel, Simpson and Beth Rosenberg, eds. *RFID: Applications, Security and Privacy.* Upper Saddle River, NJ: Addison-Wesley, 2006. Page 137.

7. Emdeon. www.emdeon.com/pdfs/Emdeon%20Script%20Flyer.pdf.

8. _____. "Consumers' Average Prescription Drug Copay Drops as Generic Drug Use Grows." *Business Wire*, April 16, 2008. Downloaded 28 August 2008 at: http://findarticles.com/p/articles/mi_m0EIN/is_2008_April_16/ai_n25156081

9. Bogdanich, Walt. "Counterfeit Drugs' Path Eased by Free Trade Zones." *New York Times*, 17 December 2007.

10. Eban, Katherine. *Dangerous Doses*. New York: Harcourt, 2005. Page 2.

11. Raynaerts, Jan-Willem. "Keeping Bogus Drugs Out of the Medicine Cabinet." *RFID Journal*, 6 August 2007.

12. Bacheldor, Beth. "Pfizer to Tag Celebrex." *RFID Journal*, November 14, 2006.

13. Meranda, Mike. "EPC Value Model: Assessing the Value of RFID Technology and EPCglobal Standards for Healthcare and Life Sciences Manufacturers." White paper, EPCGlobal, 2005. White Paper Library: *RFID Journal,*

14. Wessel, Rhea. "UK Pharmacy Test Shows Promise." *Contactless News*, 14 April 2006. Downloaded 28 August 2008 at: www.contactlessnews.com/2006/04/14/uk-pharmacy-test-shows-promise

15. Blanchard, Ben. "In Latest Scare, China Finds Fake Veterinary Drugs." Reuters, 21 June 2007. Downloaded 28 August 2008 at: uk.reuters.com/article/latestCrisis/idUKPEK27548620070621

16. Quirk, Ronald. "e-Pedigree's Evolution." *RFID Journal*, 5 March 2007.

17. Eban, Katherine, op. cit., page 16.

18. Blanchard, Ben, op. cit.

19. Quirk, Ronald, op. cit.

20. Eban, Katherine, op. cit., pp. 31–41.

21. Bogdanich, Walt, op. cit.

22. Nathanson, Drew. "Pharma to Pioneer Item-level Tracking." *RFID Journal*, 23 October 2006.

23. Brewin, Bob. "FDA Backs RFID Tags for Tracking Prescription Drugs." *Computerworld*, 23 February 2004, v. 38, issue 8, page 4.

24. _____. "Pharma Adoption Still Slow." *RFID Journal*, 23 April 2007.

25. O'Connor, Mary C. "Purdue Moving Oxycontin RFID Pilot to Full Production." *RFID Journal*, 13 February 2007.

26. O'Connor, Mary C. "EPCglobal ratifies E-Pedigree Standard." *RFID Journal*, 11 January 2006.

27. Bacheldor, Beth. "Stakes are High for Mexican Pharma RFID Mandate." *RFID Journal*, 26 September 2006.

28. Pearson, Joseph. "An RFID Tag Data Security Infrastructure Approach for Items." *RFID Journal*, 18 December 2006.

29. _____. "Protecting Against Counterfeits, Generics and Substitutes with RFID." White paper, SkyeTeck, 21 March 2007. Downloaded 28 August 2008 at: www.rfid-world.com/whitepaper/203101200

30. _____. "Improving Security with Identification Printing." White paper, Zebra Technologies. 2003. Downloaded 28 August 2008 at: www.rfidjournal.com/whitepapers/download/91

31. SkyeTech, op. cit.

32. Rasco, Barbara A. and Bledsoe, Gleyn E., op. cit.

33. O'Connor, Mary C. "Hyan and Paralec Teaming on Tags for Products, Mass Transit." *RFID Journal*, 7 September 2006.

34. Wasserman, Elizabeth. "RFID Is in Fashion." *RFID Journal*, Downloaded 28 August 2008 at www.rfid-journal.com/article/articleprint/2408/-1/359/

35. Wessel, Rhea. "RFID Ripens Cheese." *RFID Journal*, 8 January 2007.

36. Wessel, Rhea. "RFID System Opens Up Debate at Sweden's Parliament." *RFID Journal*, 18 September 2006.

37. Bacheldor, Beth. *RFID Journal*, "Missouri School District Puts RFID on Buses." *RFID Journal*, 9 November 2006.

38. SkyeTech, op. cit.

39. Shanmugam, Rajahravisankar. "PLM-RFID Combined Solutions to Solve New Business Issues." White paper, Tata Consultancy Services, Undated. Downloaded 28 August 2008 at: www.rfidjournal.com/whitepapers/4/4

40. Machrone, Bill. "RFID: Promise & Peril." *PC Magazine*, 25 November 2003, v. 22, issue 21.

41. Wessel, Rhea. "German Meat-Tracking Project Focuses on Lasers and RFID." *RFID Journal*, 2 Janurary 2007.

42. _____. "RFID Overview" White paper, Caslon Analytics, undated. www.caslon.com.au/rfidprofile.htm

43. Hofstede, Geert. *Culture's Consequences*. Thousand Oaks, CA: Sage Publications, 2001. Page 170.

44. _____. "Group Opposes Lifting Ban On US Beef Under Pressure." *Japan Today*. Downloaded 28 August 2008 at: archive.japantoday.com/jp/news/327450

45. Rasco, Barbara A. and Bledsoe, Gleyn E., op. cit.

46. Talbot, David, op. cit.

47. Gregory, Jonathan. "RFID and SAP: A Strategic Vision" White paper, Computer Sciences Corporation, May 2006. Downloaded 28 August 2008 at: www.csc.com/aboutus/leadingedgeforum/knowledgeli-brary/uploads/1128_1.pdf

48. Rasco, Barbara A and Bledsoe, Gleyn E., op. cit.

49. Rasco, Barbara A and Bledsoe, Gleyn E., op. cit.

50. Gregory, Jonathan, op. cit.

51. _____. "Advance ID's Tiny Chip Can Make a Difference in Protecting from Faulty Tire Recall System…" *Business Wire*. TMC.net. 25 February 2008. Downloaded 28 August 2008 at: www.reuters.com/article/pressRelease/idUS125546+25-Feb-2008+BW20080225

52. Rasco, Barbara A and Bledsoe, Gleyn E., op. cit.

53. Golan Elise and Krissof, Barry and Kuchler, Fred and Calvin, Linda and Nelson, Kenneth and Price, Gregory. "Traceablity in the US Food Supply: Economic Theory and Industry Studies." *Agricultural Economic Report* Number 830. United States Department of Agriculture. HD9005. (Undated, bibliography up to 2004) Page 349.

54. Ibid.

55. Rasco, Barbara A. and Bledsoe, Gleyn E., op. cit.

56. Golan et al, op. cit.

Chapter 10

1. _____. "Security of RFID Systems in Spite of Hacked Mifare-Classic RFID Chips." White paper, Feig Electronics. 23 July 2008. Downloaded 28 August 2008 at: www.rfidsolutionsonline.com/article.mvc/Security-RFID-infrastructure-0002

2. Want, Roy. *RFID Explained: A Primer on Radio Frequency Identification Technologies*. Morgan & Claypool, 2006. Page 30.

3.. Swedberg, Claire. "Amsterdam Tourists Go Contactless." *RFID Journal*, 23 March 2007.

4. Corum, Chris. "Interview: Talking Contactless and EMC with Visa's Patrick Gautier." 12 May 2005. Downloaded 28 August 2008 at:
www.secureidnews.com/2005/05/12/ interview-talking-contactless-and-emv-with-visas-patrick-gautier/

5. O'Connor, Mary C. "Chinese Railway Switching to RFID Transit Cards." *RFID Journal*, 30 August 2006.

6. Swedberg, Claire. "Moscow Metro Tries RFID-Enabled Ticketing." *RFID Journal*, 15 February 2007.

7. Williams, Andy. "NFC Goes From Pilot to Commercialized Rollout in Germany." *Contactless News*, 18 May 2006. Downloaded 28 August 2008 at:
www.secureidnews.com/2006/05/18/nfc-goes-from-pilot-to-commercialized-rollout-in-germany/

8. _____. "Contactless Payments and the Retail Point of Sale: Application, Technologies and Transaction Models." White paper, Smart Card Alliance. March 2003. Downloaded 28 August 2008 at:
www.smartcardalliance.org/pages/publications-contactless-payment-report

9. _____. "Dubai Airport Expedites Travel with Contactless Cards." *Contactless News*, 13 November 2006. Downloaded 28 August 2008 at:
www.contactlessnews.com/2006/11/13/dubai-airport-expedites-travel-with-contactless-cards

10. _____. "US Gov Hints at Major Passenger Tracking System." RFIDupdate.com, 25 April 2007. Downloaded 28 August 2008 at: www.rfidupdate.com/articles/index.php?id=1346

11. Cox, John. "Industry Group to Advance Near Field Wireless." *Network World*, 12 June 2006. Downloaded 28 August 2008 at: www.networkworld.com/news/2006/061206-near-field-wireless.html

12. Schwartz, Ephraim. "The Case for Active RFID." *INFO World*, 27 June 2005. Downloaded 28 August 2008 at: www.infoworld.com/article/05/06/21/26OPreality_1.html

13. O'Connor, Mary C. "New Approach to RFID-Powered Building Security." *RFID Journal*, 5 October 2005.

14. Baard, Mark. "RFID Invades Capital." *Wired*, 7 March 2005. Downloaded 28 August 2008 at:
www.wired.com/politics/security/news/2005/03/66801.

15. _____. "RF Ideas Enables Use of Prox and Contactless cards for Logon, Payment and Legacy ID." *Contactless News*, 24 May 2005. Downloaded 28 August 2008 at: www.contactlessnews.com/2005/05/24/rf-ideas-enables-use-of-prox-and-contactless-cards-for-logon-payment-and-legacy-id/

16. O'Connor, Mary C. "At Infosys, a Live Lab RFID App Eases Parking." *RFID Journal*, 22 February 2007.

17. Finkenzeller, Klaus. RFID *Handbook: Fundamentals and Applications in Contactless Smart Cards and Identification* (2nd ed.). Chichester: John Wiley & Sons Ltd., 2003. Page 360.

18. Swedberg, Claire. "Argentine Drivers Use Passive Tags to Pay Bridge Tolls." *RFID Journal*, 8 September 2006

19. Ferguson, Glover T. "Have Your Objects Call My Objects." *Harvard Business Review*, June 2002, v. 80, issue 6.

20. Collins, Jonathan. "At Wentworth Club, RFID Tags are Members-Only." *RFID Journal*, 25 July 2006.

21. Heinrich, Claus. Op. cit., page 49.

22. Kasavana, Michael. "Cell Phones: A Key Player in Proximity Payment Systems." AMOnline.com. 5 May 2006. www.amonline.com/print/Automatic-Merchandiser/Cell-Phones—A-Key-Player-in-Proximity-Payment-Systems/1$16096

23. O'Connor, Mary C. "Ocean City Plans to RFID-Enable its Beaches." *RFID Journal*, 26 July 2007.

24. Mullin, Bernard James and Hardy, Steven and Sutton, William Anthony. *Sport Marketing*, Champaign, IL: Human Kinetics Publishers, 2007. Page 55.

Chapter 11

1. O'Connor, Mary C. "Philly to get RFID-Enabled Vending Machines." *RFID Journal*, 28 June 2006.

2. ___. Smart Card Alliance "Contactless Payments: The Retailer Experience." Webinar. Downloaded 28 August 2008 at:
www.smartcardalliance.org/pages/activities-events-contactless-payments-retailer-webinar

3. McGrath, James C. "Micropayments: The Final Frontier for Electronic Consumer Payments." June 2006. Federal Reserve Bank of Philadelphia. Downloaded 28 August 2008 at: www.philadelphiafed.org/pcc/papers/2006/D2006JuneMicropaymentsCover.pdf

4. McGrath, James C., op. cit.

5. Will, George. "The End of Umpire?" *Washington Post*, 15 June 2008. Page B7.

6. Cohen, Alan. "Fast Food." *PC Magazine,* 4 May 2004.

7. O'Connor, Mary C. "RFID Lands a Role on Broadway." *RFID Journal,* 25 January 2007.

8. Swedberg, Claire. "North American Bancard Gives Out Contactless Payment Systems." *RFID Journal,* 20 September 2006.

9. McGrath, James C., op. cit.

10. _____. "Wireless Customer Loyalty—What's Next for RFID?" *National Petroleum News,* June 2000, v. 92, issue 6, page 64.

11. Finkenzeller, Klaus. *RFID Handbook: Fundamentals and Applications in Contactless Smart Cards and Identification,* Second Edition. Chichester: John Wiley & Sons Ltd., 2003, pp. 221–223.

12. _____. "Contactless Payments: Consumer Attitudes and Acceptance in the United States." White paper, Javelin Strategy and Research, 2006. Downloaded 28 August 2008 at: www.smartcardalliance.org/newsletter/december_2006/feature_1206.html

13. Javelin, op.cit.

14. O'Connor, Mary C. "Survey Predicts Majority of Retailers will Accept RFID Payments by Fall 2008." *RFID Journal,* 7 February 2007.

15. Swedberg, Claire. "North American Bancard Gives Out Contactless Payment Systems." *RFID Journal,* 20 September 2006.

16. _____. Contactless Payments: Delivering Merchant and Consumer Benefits." Smart Card Alliance. April 2004. Downloaded 28 August 2008 at: www.smartcardalliance.org/pages/publications-contactless-payments-benefits-report

17. _____. "Contactless Smart Cards vs. EPC Gen2 RFID Tags: Frequently Asked Questions." White paper, Smart Card Alliance, July 2006. Downloaded 28 August 2008 at: www.smartcardalliance.org/resources/pdf/EPC_Gen_2_FAQ_FINAL.pdf

18. Fabris, Nicole. "Cards Today, Phones Tomorrow: Contactless Payment's Migration to NFC Devices." ABI Research, 23 May 2006. Downloaded 28 August 2008 at: www.abiresearch.com/abiprdisplay.jsp?pressid=652

19. Kasavana, Michael. "Cell Phones: A Key Player in Proximity Payment Systems." AMonline.com. 5 May 2006. Downloaded 28 August 2008 at: www.amonline.com/print/Automatic-Merchandiser/Cell-Phones—A-Key-Player-in-Proximity-Payment-Systems/1$16096

20. O'Connor, Mary C. "NFC Scores High at Atlanta Arena." *RFID Journal,* 7 September, 2006.

21. Swedberg, Claire. "Philips, SK to Test NFC in Seoul." *RFID Journal,* 26 May 2006.

22. O'Connor, Mary C. "Discover Teaming with Motorola on NFC Banking Trial." *RFID Journal,* 14 February 2007.

23. Kasavana, Michael, op. cit.

24. Swedberg, Claire. "SnoMountain Skiers Use RFID to Play and Pay." *RFID Journal,* 30 Janurary 2007.

25. Swedberg, Claire. "Two Ohio Water Parks Become RFID-Enabled." *RFID Journal,* 14 December, 2006.

26. Javelin, op. cit.

27. Atkinson, Jimmy. "Contactless CreditCards: Consumer Report." April 2006. Downloaded 28 August 2008 at: www.findcreditcards.org/reports/contactless.html

28. _____. "EPC Gen2 RFID Tags vs. Contactless Smart Cards: Frequently Asked Questions." White paper, Smart Card Alliance. Downloaded 28 August 2008 at: www.smartcardalliance.org/pages/publications-epc-gen2-faq

29. "Contactless Payments: Delivering Merchant and Customer Benefits." Op. cit.

30. "Contactless Payments: Consumer Attitudes and Acceptance in the United States." Op. cit.

31. Kasavana, Michael, op cit.

32. Javelin, op. cit.

33. Atkinson, op cit.

34. _____. "Contactless Payment: The Top Ten Questions." Downloaded 28 August 2008 at: www.smartcardalliance.org/resources/pdf/Final_Contactless_Payment_Backgrounder.pdf

35. Wilson, Chuck. *Get Smart*. Richardson, TX: Mullaney Publishing, 2001. Pages 42–43.

36. "Contactless Payments: Delivering Merchant and Customer Benefits." Op. cit.

37. _____. "Wireless Customer Loyalty: What's Next for RFID?" *National Petroleum News,* June 2000, v. 92, issue 6, p. 64.

Chapter 12

1. Personal Interview: Steve Sabicer: Medtronic Inc.

2. Harris, Gardiner. "Report Finds a Heavy Toll from Medication Errors." *New York Times*. July 21, 2006. Downloaded 28 August 2008 at: www.nytimes.com/2006/07/21/health/21drugerrors.html.

3. Associated Press. "Drug Errors Injure More than 1.5 Million a Year." MSNBC.com, 20 July 2006, 4:51 p.m. Downloaded 28 August 2008 at: www.msnbc.msn.com/id/13954142/

4. _____. "Embedded RFID: Transforming the Management of Medical Devices and Supplies." Webinar. SkyeTech, 2007. Downloaded 28 August 2008 at: www.skyetek.com/RFIDWebinars/WebinarMedical/tabid/463/Default.aspx

5. _____. "Patient Safety Applications of Bar Code and RFID Technologies." White paper, Zebra Technologies, 2005. Downloaded 28 August 2008 at: http://downloads.racoindustries.com/downloads/RFID/PatientSafetyApplicationsWP.pdf

6. Bacheldor, Beth. "RFID-enabled Handheld Helps Nurses Verify Meds." *RFID Journal,* 10 July 2007.

7. Swedberg, Claire. "DHL Expects to Launch "Sensor Tag" Service by Midyear." *RFID Journal,* 19 January 2007.

8. Bacheldor, Beth. "InfoLogix Regenerates SurgiChip." *RFID Journal,* 5 March 2008.

9. _____. "Radio Frequency Identification and Wireless Solutions for Healthcare Service Providers." White paper, Alvin Systems, 2005. Downloaded 28 August 2008 at: www.alvinsystems.com/resources/pdf/healthcare_rfid.pdf

10. Zebra. "Patient Safety Applications…" Op. cit.

11. Bacheldor, Beth. "Siemens Launches RFID Pilot to Track Surgical Sponges, Procedures." *RFID Journal,* 24 April 2007.

12. Bacheldor, Beth. "RFID-enabled Surgical Sponges a Step Closer to OR." *RFID Journal,* 27 June 2007.

13. Collins, Jonathan. UK Agency Plans RFID Trial to Staunch Transfusion Errors." *RFID Journal,* 25 August 2006.

14. Wessel, Rhea. "RFID-enabled Locks Secure Bags of Blood." *RFID Journal,* 26 September 2006.

15. Bacheldor, Beth. "Calif Startup Develops RFID-Enabled Products to Track Medical Tests." *RFID Journal,* 14 August 2006.

16. McCoy, Jim. "Spot by Inner Wireless: a Rational Solution for Healthcare Asset Tracking." White paper, Inner Wireless, 2006. Downloaded 28 August 2008 at: www.antennasonline.com/images/Whitepapers/Innerwireless.pdf

17. McCoy, Jim, op. cit.

18. McGuinness, Michael and Rideout, Jeffrey. "Optimizing a Wireless LAN for Location." *RFID Journal,* 30 October 2006.

19. McCoy, Jim, op. cit.

20. McCoy, Jim, op. cit.

21. Wasserman, Elizabeth. "A Healthy ROI." *RFID Journal,* 15 October 2007.

22. McGuinness, Michael and Rideout, Jeffrey, op. cit.

23. Zebra. "Patient Safety Applications…" Op cit.

24. Swedberg, Claire. First Responders Can Tag Victims for Tracking." *RFID Journal,* 6 October 2006.

25. Torrieri, Marisa. "FDA-approved RFID technology eases ER visits, reduces wrong-site surgery." *Contactless News,* 22 November 2005. Downloaded 28 August 2008 at: www.contactlessnews.com/.../11/22/fdaapproved-rfid-technology-eases-er-visits-reduces-wrongsite-surgery/

26. _____. "Smart Cards in US Healthcare: Benefits for Patients, Providers and Payers." White paper, Smart Card Alliance, February 2007. Downloaded 28 August 2008 at: www.smartcardalliance.org/pages/publications-smart-cards-in-healthcare

27. Swedberg, Claire. "MedicAlert Aims to RFID-Enable Medical Records." *RFID Journal,* 16 February 2007.

28. _____. "Smart Cards in US Healthcare: Benefits for Patients, Providers and Payers." White paper, Smart Card Alliance, February 2007. Downloaded 28 August 2008 at: www.smartcardalliance.org/pages/publications-smart-cards-in-healthcare

29. Gambon, Jill. "RFID Frees Up Patient Beds." *RFID Journal,* 28 August 2006.

30. Wasserman, Elizabeth. "A Healthy ROI." *RFID Journal,* 15 October 2007.

31. Bacheldor, Beth. "Health Facility Uses RTLS to Provide Concierge Care." *RFID Journal,* 9 October 2007.

32. Skyetek. "Embedded RFID." Op. cit.

33. _____. "Smart Cards in US Healthcare…" Op. cit.

34. Collins, Jonathan. "Blood Storage Provider Plans RFID Pilot." *RFID Journal,* 7 July 2006.

35. Fishkin, Kenneth and Lundell, Jay. "RFID in Healthcare" In Garfinkel, Simpson and Beth Rosenberg, eds. RFID: Applications, Security and Privacy. Upper Saddle River, NJ: Addison-Wesley, 2006. Page 224.

36. Fishkin, Kenneth and Lundell, Jay. Op. cit. Page 224.

37. Pyke, Bob Jr. "Achieving Global Telehealth: Words from the Trenches: A talk with David Balch." Downloaded 28 August 2008 at: www.telehealth.net/interviews/balch.html

38. Gregory, Jonathan. "RFID and SAP: A Strategic Vision." White paper, Computer Sciences Corporation, May 2006. Downloaded 28 August 2008 at: www.csc.com/aboutus/leadingedgeforum/knowledgelibrary/uploads/1128_1.pdf

39. Finkenzeller, Klaus. *RFID Handbook: Fundamentals and Applications in Contactless Smart Cards and Identification,* (2nd ed.). Chichester: John Wiley & Sons Ltd., 2003. Pages 392–393.

40. Swedberg, Claire. "Researchers Develop RFIFD System to Monitor Acid Reflux." *RFID Journal,* 10 April 2007.

41. _____. "RFID Invention to Detect Esophageal Reflux." RFIDupdate.com, 1 May 2007. Downloaded 28 August 2008 at: www.rfidupdate.com/articles/index.php?id=1371

42. _____. "Radio Frequency Patient Monitoring: A Cost-Effective patient/Staff model for the Emergency Department." White paper, RFT Technologies, Undated. Downloaded 28 August 2008 at: www.rft.com/templatefiles/includes/common/displayFile.ashx?fileId=197

43. Bacheldor, Beth. "New System Reports Patient Falls." *RFID Journal,* 28 June 2006.

44. Smith, Joshua R, and Fishkin, Kenneth, and Jiang, Biang, et al. "RFID-based Techniques for Human Activity Detection." *Communications of the ACM,* September 2005. V. 48, N. 9, pp 39-44.

45. Bacheldor, Beth. "RFID Fills Security Gap at Psychiatric Ward." *RFID Journal,* 24 October 2006.

46. Simpson, Garfinkel and Beth Rosenberg, editors. *RFID: Applications, Security and Privacy,* pp. 216-221. New York: Addison-Wesley, 2006.

47. Rangarajan, T.S. and Vijaykumar, Anand. "RFID in Clinical Trials." White paper, Tata Consultancy Services, undated. Downloaded 28 August 2008 at: www.tcs.com/SiteCollectionDocuments/White%20Papers/RFID%20in%20Clinical%20Trials.pdf

48. Rangarajan, T.S. and Vijaykumar, Anand. Op. cit.

49. Rangarajan, T.S. and Vijaykumar, Anand. Op. cit.

Appendix

1. Gregory, Jonathan. "RFID and SAP: A Strategic Vision." White paper, Computer Sciences Association, May 2006.

2. Wilson, Chuck. *Get Smart*, page 13. Richardson, TX: Mullaney Publishing Group. 2001.

Selected Bibliography

Anderson, Stacy. "Passport Chip Sets Security Concerns." *Wall Street Journal*, 9 August 2006.

Ashton, Kevin. "Do We Need Digital Turtles?" *RFID Journal*, 19 February 2007.

_____. "Privacy Is Simple" *RFID Journal*, Undated.

Associated Press. "Drug Errors Injure More Than 1.5 Million a Year." MSNBC.com, 20 July 2006, 4:51 PM.

Atkinson, Jimmy. "Contactless Credit Cards Consumer Report." Findcreditcards.org, 3 April 2006.

Atkinson, Rob. "RFID: There's Nothing to Fear Except Fear Itself." Presented at 16th Computers, Freedom and Privacy Conference, 15 May 2006. The Information and Technology Foundation.

Bacheldor, Beth. "GE Sensing, Dust Networks to Develop Wireless Sensors." *RFID Journal*, 15 May 2007.

_____. "RFID Enters the Pig Pen." *RFID Journal*, 1 February 2007.

Barnes, James G. *Secrets of Customer Relationship Marketing*. New York: McGraw-Hill, 2001.

Baylor, Coreen. "Ten Technologies That Are Reinventing the CRM Industry." *Customer Relationship Management*, December 2004, V. 8. N. 12, pp. 44–50.

Bergeron, Brian. *The Eternal E-Customer*. New York: McGraw-Hill, 2001.

Berson, Alex, Stephen Smith, and Kurt Thearling. *Building Datamining Applications for CRM*. New York: McGraw-Hill, 2000.

Bliss, Jeanne. *"Customers Defect When the Silos Don't Connect." DestinationaCRM.com, 1 August 2006.*

Boriello, Gaetano. "RFID: Tagging the World." *Communications of the ACM*, September 2005, V. 48, N. 9.

Brown, Stanley A. *Customer Relationship Management: A Strategic Imperative in the World of E-Business*. Toronto: John Wiley & Sons, 2000.

Buttle, Francis. *Customer Relationship Management: Concepts and Tools*. Burlington, MA: Elsevier Butterworth Heineman, 2004.

_____. "The S.C.O.P.E. of Customer Relationship Management." White Paper. *International Journal of Customer Relationship Management*, January 1999, pp. 327–336.

Chopra, Sunil and Peter Meindl. *Supply Chain Management: Strategy, Planning and Operation*. Saddle River, NJ: Prentice-Hall, 2001.

Clarke, R.A. "Human Identification in Information Systems: Management Challenges and Public Policy Issues." *Information Technology and People*, December 1994, V. 7 N. 4, pp. 6–37. http://www.anu.edu.au/people/Roger.Clarke/DV/HumanID.html

Clegg, Brian. *Capturing Customers' Hearts*. Saddle River, NJ: Prentice-Hall, 2000.

Collins, Jonathan. "Blood Storage Provider Plans RFID Pilot." *RFID Journal*, 7 July 2006.

_____. "Hertz Trial Uses RFID Cards Instead of Keys." *RFID Journal*, 9 August 2006.

Cram, Tony. *Customers That Count: How to Build Living Relationships with Your Most Valuable Customers*. London: Pearson/Prentice Hall, 2001.

Crane, Michael. "Intelligent Network Services Can Make RFID More Productive." *RFID Journal*, 11 December 2006.

Culler, David E. "Smart Sensors to Network the World." *Scientific American*, June 2004. http://www.intel.com/research/exploratory/smartnetworks.htm

Cusack, Michael. *Online Customer Care: Strategies for Call Center Excellence*. Milwaukee: ASQ Quality Press, 1998.

_____. "The Use of RFID for Human Identity Verification." RFID Subcommittee Report. Dept of Homeland Security. Draft: 12/6/2006.
http://www.dhs.gov/xlibrary/assets/privacy/privacy_advcom_12-2006_rpt_RFID.pdf

Eban, Katherine. *Dangerous Doses: How Counterfeiters Are Contaminating America's Drug Supply.* Orlando: Harcourt, 2005.

Eckfield, Bruce. "Change Is in the air." *The Economist,* 12 March 2005, V. 374, Issue 8417.

_____. "The Future Is Still Smart." *The Economist,* 26 June 2004, V. 371, Issue 8381.

_____. "What does RFID do for the Consumer." *Communications of the ACM,* September 2005, V. 48. N 9.

_____. "Where's the Smart Money?" *The Economist,* 9 February 2002, V. 362, Issue 8259.

Embry, Wayne. "Are You Ready for RFID?" White Paper. SAS, Undated.
www.sas.com/ci/whitepapers/102274.pdf

Evans, David S., and Richard Schmalensee. *Paying with Plastic.* 2nd ed. Cambridge, MA: MIT Press, 2005.

Ferguson, Glover T. "Have Your Objects Call My Objects." *Harvard Business Review,* June 2002, V. 80, Issue 6.

Ferreira, Devon, and Girish A. Ramachandran. "Simplify: Leveraging RFID and Other Pervasive Technologies to Achieve Transaction Efficiency in Customer Retailing." White Paper. Infosys, April 2006.
http://www.infosys.com/RFID/simplify-consumer-retailing.pdf

Fildes, Jonathan. "Chips With Everything." *New Scientist,* 19 October 2002, V. 176, Issue 2365.

Finkenzeller, Klaus. *RFID Handbook: Fundamentals and Applications in Contactless Smart Cards and Identification.* 2nd ed. Chichester: John Wiley & Sons, Ltd., 2003.

Floerkemeir, Christian, Roland Schneider, and Marc Langheninreich. "Scanning with a Purpose: Supporting the Fair Information Principles in RFID Protocols." *Revised Selected Papers from the Second International Symposium on Ubiquitous Computing Systems* (UCS 2004), November 8–9, 2004. Ubiquitous Computing Systems.
http://www.vs.inf.ethz.ch/publ/papers/floerkem2004-rfidprivacy.pdf

Foss, Bryan, and Merlin Stone. *CRM in Financial Services: A Practical Guide to Making Customer Relationship Management Work.* London; Milford, CT: Kogan Page, 2002.

Freeland, John G. *The Ultimate CRM Handbook.* New York: McGraw-Hill, 2003.

Gamble, Paul, Merlin Stone, and Neil Woodcock. *Up Close and Personal?: Customer Relationship Marketing @ Work.* London; Philadelphia: Kogan Page, 2006.

Gambon, Jill. "Florida to Require RFID Tagging for Exotic Pets." *RFID Journal,* 12 February 2007.

Garfinkel, Simpson. "An RFID Bill of Rights." *Technology Review,* October 2002, V. 105, Issue 8.

Garfinkel, Simpson, and Beth Rosenberg, eds. *RFID: Applications, Security and Privacy.* Upper Saddle River, NJ: Addison-Wesley, 2006.

Glover, Bill, and Bhatt, Himanshu. *RFID Essentials.* Sebastopol, CA: O'Reilly Media Inc., 2006.

Goldenberg, Barton J. *CRM Automation.* Saddle River, NJ: Prentice-Hall, 2002.

Greenberg, Paul. *CRM at the Speed of Light.* 2nd ed. New York: McGraw-Hill, 2002.

Greenfield, Adam. *Everyware: The Dawning Age of Ubiquitous Computing.* Berkeley, CA: New Riders (Pearson Education), 2006.

Greengard, Samuel. "Mississippi Blood Services Banks on RFID." *RFID Journal,* 7 August 2006.

Gregory, Jonathan. "RFID and SAP: A Strategic Vision." White Paper. Computer Sciences Corporation, May 2006. http://www.csc.com/aboutus/leadingedgeforum/knowledgelibrary/uploads/1128_1.pdf

Griffin, Jill, and Michael W. Lowenstein. *Customer Winback: How to Recapture Lost Customers and Keep Them Loyal.* San Francisco: Jossey Bass Publishing. 2001.

Gunther, Oliver, and Sarah Spickermann. "RFID and the Perception of Control: The Consumer's View." *Communications of the ACM,* September 2005, V. 48 N. 9, pp 73–76.

Gupta, Puneet. "RFIDs: Enabling Sense and Respond Businesses." *Express Computer India.*
http://www.expresscomputeronline.com/20031222/technology.shtml

Hanebeck, Christian. "Building the Intelligent Network of Collected Items" (#4 of 6). White Paper. Globeranger, 2006.

_____. "Managing Data from RFID and Sensor-Based Networks" (#2 of 6). White Paper. Globeranger, 2006.

_____. "Processes Management and RFID" (#3 of 6). White Paper. Globeranger, 2006.

Harper, Jim. *Identity Crisis: How Identification Is Overused and Misunderstood*. Washington, D.C.: Cato Institute, 2006.

Harris, Gardiner. "Report Finds a Heavy Toll from Medication Errors." *New York Times*, 21 July 2006.

Heinrich, Claus. *RFID and Beyond*. Indianapolis: Wiley Publishing, Inc., 2005.

Hunter, Richard. *World Without Secrets: Business, Crime and Privacy in the Age of Ubiquitous Computing*. New York: John Wiley and Sons, 1998.

Hutto, Julie, and Robert D. Atkinson. "Item-Level RFID Technology Redefines Retail Operations with Real-Time, Collaborative Capabilities." White Paper. IBM, 2004. www-03.ibm.com/industries/retail/doc/content/bin/rfid_redefine_1.pdf

_____. "Near Field Communication in the Real World." White Paper. Innovision, Undated. http://www.nfc-forum.org/resources/white_papers/Innovision_whitePaper1.pdf

_____. "Near Field Communication in the Real World Part II: Using the Right NFC Tag Type for the Right NFC Application." White Paper. Innovision, Undated. www.nfc-forum.org/resources/white_papers/Innovision_whitePaper2.pdf

_____. "Radio Frequency Identification: Little Devices Making Big Waves." Progressive Policy Institute. *Policy Report*, October 2004.

_____. "The Role of Chips in Making Identification Documents More Secure." White Paper. Infeon Technologies, 14 July 2005.

Jaffe, Joseph. *Life After the Thirty Second Spot: Energize Your Brand With a Bold Mix of Alternatives to Traditional Advertising*. Hoboken, NJ: John Wiley & Sons, 2005.

Jaselskis, Edward J., and Tarek El Misalami. "Contactless Payments: Consumer Attitudes and Acceptance in the United States. White Paper. Javelin Strategy and Research, 2006. www.smartcardalliance.org/pages/publications-contactless-payments-attitudes-acceptance

_____. "Implementing Radio Frequency Identification in the Construction Process." *Journal of Construction Engineering and Management*, November/December 2003, V. 129. I. 6, pp. 680–688.

Kahn, Barbara E., and Leigh McAlister. *Grocery Revolution: The New Focus on the Consumer*. New York. Addison-Wesley. 1997.

Kasanoff, Bruce. *Making It Personal: How to Profit From Personalization Without Invading Privacy*. Cambridge, MA: Perseus Publishing, 2001

Kasavana, Michael. "Cell Phones: A Key Player in Proximity Payment Systems." AMOnline.com, 5 May 2006. www.amonline.com/publication/article.jsp?pubId=1&id=16096

Kim, Sung Woo, et al. "Sensible Appliances: Applying Context Awareness to Appliance Design." *Journal of Personal and Ubiquitous Computing*, 15 February 2004, V. 8, p. 184–191.

Knox, Simon, Stan Maklan, Adrian Payne, Joe Peppard, and Lynette Ryals. *Customer Relationship Management: Perspectives From the Marketplace*. New York: Butterworth Heinemann, 2003.

Laursen, Wendy. "Managing the MegaFlock." *IEEE Review*, February 2006, pp 38–42.

Liautaud, Bernard. *E-Business Intelligence: Turning Information Into Knowledge Into Profit*. New York: McGraw-Hill, 2001.

Luedtke, Joe. "Toward Pervasive Computing: RFID Tags: Pervasive Computing in Your Pocket, on Your Key Chain and in Your Car." *DM Review*, July 2003.

Margulius, David L. "The Rush to RFID." *Info World*, 12 April 2004.

Meranda, Mike. "EPC Value Model: Assessing the Value of RFID Technology and EPCglobal Standards for Healthcare and Life Sciences Manufacturers." White Paper. EPCglobal, 2005. www.whitepapers.silicon.com/0,39024759,60164102p,00.htm

Morrell, John. "Complex Event Processing and RFID." *RFID Journal*, 5 February 2007.

Neumann, Peter G., and Lauren Weinstein. "Risks of RFID." *Communications of the ACM*, May 2006, V 49, Issue 5, p. 136.

Newell, Frederick. *Loyalty.com: Customer Relationship Management in the New Era of Internet Marketing*. New York: McGraw-Hill, 2000.

_____. *Why CRM Doesn't Work: How to Win by Letting Customers Manage the Relationship*. Princeton, NJ: Bloomberg Press, 2003.

Newell, Frederick, and Katherine Newell Lemon. *Wireless Rules: New Marketing Strategies for Customer Relationship Management Anytime, Anywhere*. New York: McGraw Hill, 2001.

Nohara, Yasunobu, Suzo Inoue, Kensuki Baba, and Hiroto Yasuura. "Quantitative Evaluation of Unlinkable ID Matching Schemes." *Proceedings of the 2005 ACM Workshop on Privacy in the Electronic Society*, 2005, p.55.

_____. "Real-Time Promotion Execution." White Paper. Oat Systems, 2005.

O'Connor, Mary C. "HP Spots New Opportunities for Passive RFID." *RFID Journal*, 18 July 2006.

_____. "RFID Cures Concrete." *RFID Journal*, 30 October 2006.

Ohkubo, Miyako, Koutarou Suzuki, and Shingo Kinoshita. "Customer Process Management: The Real-Time Enterprise Depends On the Merging of CRM and BPM." White Paper. ONYX, 2005. www.viewer.bitpipe.com/viewer/viewDocument.do?accessId=8045099

_____. "RFID Privacy Issues and Technical Challenges." *Communications of the ACM*, September 2005, V. 48, N. 9, pp. 66–71.

Payne, Adrian, and Pennie Frow. "A strategic framework for customer relationship management." *Journal of Marketing*, V. 69, N. 4.

Peel, Jeffrey. *CRM: Redefining Customer Relationship Management*. Amsterdam; Boston: Digital Press, 2002.

Peppers, Don, and Martha Rogers. "Anonymization Offers New Approach to Data Sharing." 1:1 Privacy, 14 September 2006.

_____. *Enterprise One to One: Tools for Competing in the Interactive Age*. New York: Doubleday, 1997.

_____. *Managing Customer Relationships: A Strategic Framework*. Hoboken, NJ: John Wiley & Sons, 2004.

_____. *One to One, B2B: Customer Development Strategies for the Business to Business World*. New York: Doubleday, 2001.

_____. *The One to One Future: Building Relationships One Customer at a Time*. New York: Doubleday, 1993.

_____. *The One to One Manager: Real-World Lessons in Customer Relationship Management*. New York: Doubleday, 1999.

Pering, Trevor, Rafael Ballagas, and Roy Want. "Spontaneous Marriages of Mobile Devices and Interactive Spaces." *Communications of the ACM*, September 2005, V. 48. N. 9, pp. 53–59.

Poirier, Charles, and Duncan McCollum. *RFID: Strategic Implementation and ROI*. Fort Lauderdale, FL: J. Ross & Company Publishing, 2006.

Rangarajan, T.S., and Anand Vijaykumar. "RFID in Clinical Trials." White Paper. Tata Consultancy Services, Undated. www.tcs.com/SiteCollectionDocuments/White%20Papers/RFID%20in%20Clinical%20Trials.pdf

Rangarajan, T.S., Anand Vijakumar, and Suraj S. Subramaniam. "Enabling Traceability: RFID." White paper. Tata Consulting Services, 2005. www.tcs.com/SiteCollectionDocuments/White%20Papers/Enabling%20Traceability.pdf

_____. "Upstreaming RFID: Beyond Tags and Readers." White Paper. Tata Consulting Services, 2005. www.tcs.com/SiteCollectionDocuments/White%20Papers/Up%20streaming%20RFID.pdf

Rasco, Barbara A., and Gleyn E. Bledsoe. *Bioterrorism and Food Safety*. Boca Raton: CRC Press, 2005.

Reichheld, Frederick F., and Thomas Teal. *The Loyalty Effect*. Boston: Harvard Business School Press, 1996.

Saar, Steven and Valerie Thomas. "Toward Trash That Thinks: Product Tags for Environmental Management." *Journal of Industrial Ecology*, 2003, V. 6. N. 2, pp. 133–146.

Scheneier, Bruce. *Secrets and Lies: Digital Security in a Networked World*. New York: John Wiley & Sons, 2000.

Schmitt, Bernd H. *Customer Experience Management*. New York: John Wiley & Company, 2001.

Shanmugam, Rajaravisankar. "RFID—Tackling Product Lifecycle Management Issues." White Paper. Tata Consultancy Services, September 2004. See White Paper Library: *RFID Journal*.

Shepard, Steven. "Contactless Payments and the Retail Point of Sale: Applications, Technologies and Transaction Models." White Paper. *Smart Card Alliance*, March 2003.

_____. "Logical Access Security: The Role of Smart Cards in Strong Authentication." White Paper. *Smart Card Alliance*, October 2004.
www.smartcardalliance.org/pages/publications-logical-access-report

_____. "Protecting Against Counterfeits, Generics and Substitutes with RFID." White Paper. Skyetek, 21 March 2007. www.skyetek.com/Company/NewsEvents/RFIDWhitePapers/tabid/464/Default.aspx

_____. "Proximity Mobile Payments: Leveraging NFC and the Contactless Financial Payments Infrastructure." White Paper. *Smart Card Alliance*, September 2007.
www.smartcardalliance.org/pages/publications-proximity-mobile-payments

_____. *RFID: Radio Frequency Identification*. New York: McGraw-Hill, 2005.

_____. "Smart Cards in US Healthcare: Benefits for Patients, Providers and Payers." White Paper. *Smart Card Alliance*, February 2007.
www.smartcardalliance.org/pages/publications-smart-cards-in-healthcare

Smith, Joshua R., Kenneth Fishkin, Biang Jiang, Alexander Mamishev, et al. "RFID-Based Techniques for Human Activity Detection." *Communications of the ACM*, September 2005, V. 48. N 9., pp. 39–44.

Solove, Daniel J. *The Digital Person: Technology and Privacy in the Information Age*. New York: New York University Press, 2004.

Songini, Mark. "Colorado Hopes RFID Can Protect Elk Herds." *Computerworld*, 16 January 2006.

Swedberg, Claire. "First Responders Can Tag Victims for Tracking." *RFID Journal*, 6 October 2006.

_____. "MedicAlert Aims to RFID-Enable Medical Records." *RFID Journal*, 16 February 2007.

_____. "OATSystems Launches Solutions for Tracking In-Store Product Promotions." *RFID Journal*, 21 March 2007.

Swift, Ronald S. *Accelerating Customer Relationships: Using CRM and Relationship Technologies*. Saddle River, NJ: Prentice-Hall, 2001.

Talbot, David. "Where's the Beef From?" *Technology Review*, June 2004, pp. 48–56.

Thillairajah, Velan, Sanjay Gosain, and Dave Clarke. "Realizing the Promise of RFID: Insights from Early Adopters and the Future Potential." White Paper. EAI Technologies. See White Paper Library, *RFID Journal*.

Thompson, Bob. "Customer Experience Management: Accelerating Business Performance." White Paper, part two. Customerthink Corporation and Rightnow Technologies.
www.retaintogain.com/pdf/customer_exp2.pdf

_____. "Customer Experience Management: The Value of 'Moments of Truth.'" White Paper, part one. Crmguru.com and Rightnow Technologies, 2006.
crm.rightnow.com/cgi-bin/rightnow.cfg/php/enduser/doc_serve.php?2=WPCRM-FORM-CRMX-change-TextAd-070903-CEMI

_____. "The Loyalty Connection: Secrets to Customer Retention and Increased Profits." White Paper. Rightnow Technologies, 2005.   www.rightnow.com/resource/RN_LoyaltyCRMGuru.php

Transit Cooperative Research Program: Report 108. "Car Sharing: Where and How It Succeeds." Transportation Research Board. Sponsored by the Federal Transit Administration.
www.trb.org/news/blurb_detail.asp?id=5634

Troyk, Philip. "Injectable Electronic Identification Monitoring and Stimulation Systems." *Annual Review of Biomedical Engineering*, 1999, V. 1, pp. 177–209.

Underhill, Paco. *Why We Buy: The Science of Shopping*. New York: Simon & Schuster. 1999.

Want, Roy. "RFID: A Key to Automating Everything." *Scientific American*, January 2004, V. 290, Issue 1.

_____. *RFID Explained: A Primer on Radio Frequency Identification Technologies*. San Francisco: Morgan & Claypool Publishers, 2006.

Watters, D.G., P. Jayaweera, A.J. Behr, and D.L. Herestis. "Design and Performance of Wireless Sensors for Structural Health Monitoring." *CP615 Review of Quantitative Non-Destructive Evaluation*, V. 21, 2002, American Institute of Physics.

Wessel, Rhea. "BRIDGE Project Members Press Ahead." *RFID Journal*, 27 October 2006.

_____. "German Meat-Tracking Project Focuses on Lasers and RFID." *RFID Journal*, 2 January 2007.

Wilson, Chuck. *Get Smart: The Emergence of Smart Cards in the United States and Their Pivotal Role in Internet Commerce*. Richardson, TX: Mullaney Publishing Group, 2001.

_____. "Patient Safety Applications of Bar Code and RFID Technologies." White Paper. Zebra Technologies, 2005.

Index

profitable customers, 19, 26
promotion execution, 113–116
Purdue, 152

R

Radio Frequency Identification (RFID)
  adding value, 10, 119, 134
  costs of, 4, 123
  database defenses, 67
  definition, 1
  encryption, 67
  limitations, 49–50, 123–124
  middleware, 4, 10, 96
  passwords, 66
  reader, 4, 43, 66, 111, 187
  sensors, 4, 7, 118, 136
  tag blocking, 66
  tag killing, 65–66
  tags, 4, 6, 42, 96–97, 117–119, 123
  turning off, 50, 67
  unlinkability, 67
Real-Time Location Systems (RTLS), 79, 216
real-time visibility, 4–5, 36–37, 78, 95, 117,
  128–129
recognition of customers, 36
Recycle Bank, 132
recycling, 128–133
Red Bag waste, 133
relationship control, 34, 49
relationship marketing, 19–20, 31, 41, 134–135
  benefits to customers, 20, 21, 38, 73, 119–120
  commitment, 20
  customer behavior, 20, 26–27
  customer preference, 19, 26
  customer value, 19
  trust, 39
Renaissance Fairs, 118
responsive replenishment (restocking) , 2,
  10–11, 117
retail, 97–124
  apparel retailers, 108–110
  benefits of RFID, 117–122
  interactive point of purchase, 97–98
  limitations, 123–124
  promotion deployment, 115–116
  real-time retailing, 101–102
  relationship retailing, 99–101

supply chain efficiency, 136
  retail shrink, 2
    eliminating, 122
reverse logistics, 156
RFID Customer Bill of Rights, 69–70, 213
RFID mis-use, 71–72
RFID passport, 59–60
Rice Krispies, 9
Rights of marketers, 70–71
Ritz-Carlton Hotel, 8
Rogers, Martha, 8, 33
Royal Bank of Canada, 32

S

S2CGlobal, 130
Saar and Thomas, 130
safety inventory, 2
Samsung, 189
SAP, 81
search engine, 1
Secrets of Customer Relationship Marketing
(book), 22
Seguro Popular, 152
self-service, 5, 28, 47, 99
sense-and-respond, 102-108, 136
  electronic shelf label, 107
  kiosks, 104
  mobile shopping assistants, 104–105
  monitoring gate, 105
  smart point of sale, 107
  smart shelf, 106, 110–111, 113
  smart table, 111
Shanghai Post Office, 6
shelf life, 96
silos of control, 35
silos of ignorance, 35
skimming, 55
Smart Medical Technologies, 206
smart poster, 119–120
South Korea, 119, 129, 189
Stanford/WaPo/ABC poll, 127–128
Starbucks, 183
storekeeper relationship, 44
Sudden Infant Death Syndrome, 214
supply chain technology, 10, 12, 47

Practical Books for Smart Marketers from PMP

Market Research

The 4Cs of Truth in Communication:
*How to Identify, Discuss, Evaluate and Present Stand-out,
Effective Communication*

The Art of Strategic Listening:
Finding Market Intelligence through Blogs and other Social Media

Consumer Insights 2.0:
How Smart Companies Apply Customer Knowledge to the Bottom Line

Dominators, Cynics, and Wallflowers:
Practical Strategies for Moderating Meaningful Focus Groups

Moderating to the Max!
A Full-Tilt Guide to Creative, Insightful Focus Groups and Depth Interviews

The Mirrored Window: *Focus Groups from a Moderator's Point of View*

Religion in a Free Market: *Religious and Non-Religious Americans—
Who, What, Why, Where*

Why People Buy Things They Don't Need

Mature Market/ Baby Boomers

Advertising to Baby Boomers

After Fifty: *How the Baby Boom Will Redefine the Mature Market*

After Sixty: *Marketing to Baby Boomers Reaching Their Big Transition Years*

Baby Boomers and Their Parents:
Surprising Findings about Their Lifestyles, Mindsets, and Well-Being

Marketing to Leading-Edge Baby Boomers

The Boomer Heartbeat:
Capturing the Heartbeat of the Baby Boomers Now and in the Future

Multicultural

Beyond Bodegas: *Developing a Retail Relationship with Hispanic Customers*

Hispanic Marketing Grows Up: *Exploring Perceptions and Facing Realities*

Hispanic Customers for Life: *A Fresh Look at Acculturation*

India Business: *Finding Opportunities in this Big Emerging Market*

Latinization: *How Latino Culture Is Transforming the U.S.*

Marketing to American Latinos: *A Guide to the In-Culture Approach, Part I & II*

The Whole Enchilada: *Hispanic Marketing 101*

What's Black About It?
Insights to Increase Your Share of a Changing African-American Market

Youth Markets

The Kids Market: *Myths & Realities*

Marketing to the New Super Consumer: Mom & Kid

The Great Tween Buying Machine: *Marketing to Today's Tweens*

Marketing Strategy/Management

A Clear Eye for Branding: *Straight Talk on Today's Most Powerful Business Concept*

A Knight's Code of Business
How to Achieve Character and Competence in the Corporate World

Beyond the Mission: Statement *Why Cause-Based Communications Lead to True Success*

Brand Busters: *7 Common Mistakes Marketers Make*

Marketing Insights to Help Your Business Grow

Outsmart the MBA Clones
The Alternative Guide to Competitive Strategy, Marketing, and Branding

RFID: *Improving the Customer Experience*

About the Author

MICKEY BRAZEAL is a professor in the Graduate School of Integrated Marketing Communication at Roosevelt University in Chicago.

He worked for 28 years as an advertising agency creative—the last ten as Executive Creative Director of a large Chicago agency. He has developed national TV campaigns for bicycles and banks, car waxes and corn herbicides; shampoos, spice blends and stock options; fruit juices and food stores; deodorants and drugstores; real estate companies and rental cars,; toys and tractors; and medicines among others.

His creative awards include Addies, Tellies, Eagles, Towers, Mobius Awards, and the *Gallagher Report's* Broken Pencil Award for "the most obnoxious TV commercial we've seen this month."

He has also worked in print advertising, Internet marketing, direct marketing and sales promotion. During the dotcom madness, he helped to launch several Internet businesses, some of which, including stamps.com, still thrive today.

Mickey wrote "Green Revolution: RFID and the Rise of Convenient Sustainability" for *RFID World 2008*. He wrote "Varieties of Efficiency: How RFID Asset Tracking Changes the Way Businesses Work" for the book *The Culture of Efficiency*.

Before joining the Roosevelt faculty, Mickey taught at Northwestern University and at the Stuart Graduate School of Business at IIT. He has introduced new courses in Customer Relationship Management at IIT and at Roosevelt. He is a frequent speaker on marketing and relationship management issues.